W9-AJS-676

AUSTRALIA in Pictures

Ann Kerns

Lerner Publications Company

Contents

Lerner Publishing Group realizes that current information and statistics quickly become out of date. To extend the usefulness of the Visual Geography Series, we developed www.vgsbooks.com, a website offering links to up-to-date information, as well as in-depth material, on a wide variety of subjects. All of the websites listed on www.vgsbooks.com have been carefully selected by researchers at Lerner Publishing Group. However, Lerner Publishing Group is not responsible for the accuracy or suitability of the material on any website other than <www.lernerbooks.com>. It is recommended that students using the Internet be supervised by a parent or teacher. Links on www.vgsbooks.com will be regularly reviewed and updated as needed.

INTRODUCTION 4

THE LAND 8

► Topography. Water Resources. Climate. Flora and Fauna. Natural Resources. Major Urban Centers. Secondary Cities.

HISTORY AND GOVERNMENT 20

► The First Australians. European Arrival. British Settlement. Early Reforms. Exploration and Expansion. The Gold Rush. Forming One Nation. The World Wars. Postwar Growth and Prosperity. Decades of Change. New Challenges. Government.

THE PEOPLE 40

► Ethnic Mixture. Standard of Living. Education. Health and Social Services.

Website address: www.lernerbooks.com

Lerner Publications Company
A division of Lerner Publishing Group
241 First Avenue North
Minneapolis, MN 55401 U.S.A.

web enhanced @ www.vgsbooks.com

CULTURAL LIFE 46

► Religion, Holidays, and Festivals. Literature. Art, Movies, and Music. Food. Sports and Recreation.

THE ECONOMY 56

► Mining and Manufacturing. Agriculture. Fishing and Forestry. Transportation. Trade and Tourism. The Future.

FOR MORE INFORMATION

► Timeline 66
► Fast Facts 68
► Currency 68
► Flag 69
► National Anthem 69
► Famous People 70
► Sights to See 72
► Glossary 73
► Selected Bibliography 74
► Further Reading and Websites 76
► Index 78

Library of Congress Cataloging-in-Publication Data

Kerns, Ann, 1959-
Australia in pictures / by Ann Kerns.— Rev. and updated ed.
p. cm. — (Visual geography series)
Rev. ed. of: Australia in pictures / prepared by Geography Dept., Lerner Publications Company. © 1990.
Summary: A guide to the history, government, people, culture, and economy of Australia.
Includes bibliographical references and index.
ISBN: 0-8225-0932-6 (lib. bdg. : alk. paper)
1. Australia—Juvenile literature. 2. Australia—Pictorial works—Juvenile literature. [1. Australia.]
I. Australia in pictures. II. Title. III. Visual geography series (Minneapolis, Minn.)
DU96.K47 2004
994—dc21 2003001627

Manufactured in the United States of America
1 2 3 4 5 6 - JR - 09 08 07 06 05 04

INTRODUCTION

Australia, a large country in the South Pacific Ocean, occupies the world's smallest, flattest, and driest continent. Scientists believe that Australia was once part of a huge landmass. Millions of years ago, it broke off and began drifting northward toward the equator. In this remote land, many unusual plants and animals evolved. The first people to live there remained isolated for thousands of years until explorers from Europe arrived in modern times. This contrast of ancient and new, strange and familiar, makes Australia a unique country.

At least forty thousand years ago, the Aborigines—Australia's first people—began to arrive from neighboring islands. As centuries passed, more than five hundred Aboriginal groups developed, each occupying a separate territory. A complex religion and strong patterns of social conduct governed the Aborigines' lives. Hundreds of generations of Aborigines lived undisturbed by foreign influences until the late eighteenth century, when Great Britain claimed Australia as its territory.

Britain wanted a place to which it could send criminals, and Australia met that need. As their familiarity with the remote continent increased, however, the British came to recognize Australia's economic potential. Settlers pushed Aborigines off the best land, and Australia began to develop as a European outpost in the South Pacific.

Eventually, six British colonies formed in Australia, and they often engaged in rivalry rather than in cooperation. In the late nineteenth century, the colonial governments decided unity would help them solve their common problems. The country peacefully gained its independence in 1901, while maintaining close political and commercial ties with Britain.

By the 1950s, most Australians were enjoying a high standard of living. Discoveries of large mineral deposits and energy resources added to the country's prosperity. As mining and manufacturing grew, labor shortages developed. The government encouraged immigration to provide more workers for industry.

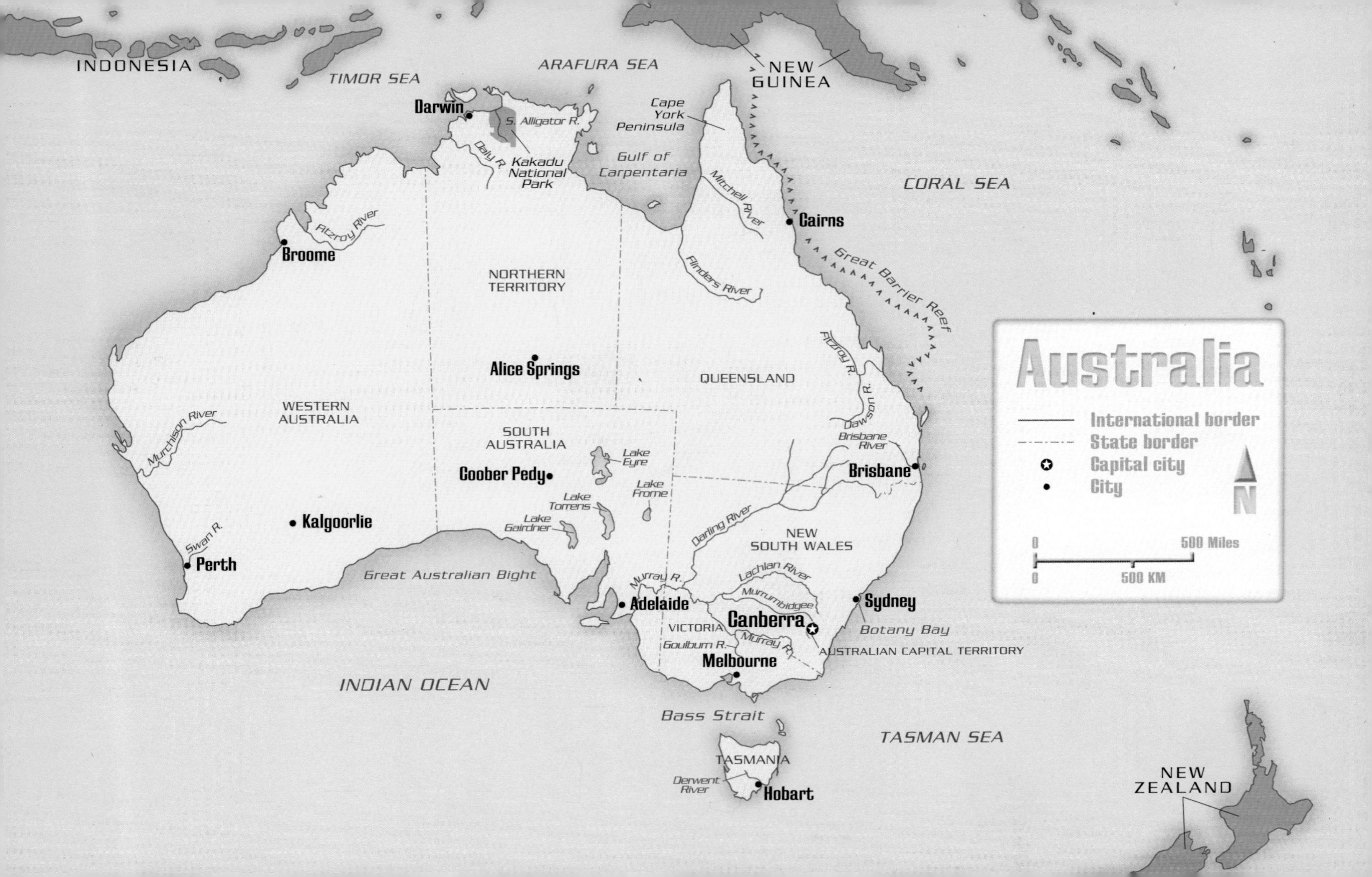

Australia
International border
State border
Capital city
City
N
0
500 Miles
0
500 KM
INDONESIA
TIMOR SEA
ARAFURA SEA
NEW GUINEA
CORAL SEA
Darwin
S. Alligator R.
Daly R.
Kakadu National Park
Cape York Peninsula
Gulf of Carpentaria
Mitchell River
Cairns
Great Barrier Reef
Fitzroy River
Broome
Flinders River
NORTHERN TERRITORY
Alice Springs
QUEENSLAND
Fitzroy R.
Dawson R.
Brisbane River
Brisbane
WESTERN AUSTRALIA
Murchison River
SOUTH AUSTRALIA
Lake Eyre
Coober Pedy
Lake Frome
Lake Torrens
Lake Gairdner
Kalgoorlie
Swan R.
Perth
Darling River
NEW SOUTH WALES
Great Australian Bight
Murray R.
Lachlan River
Murrumbidgee
Adelaide
Sydney
Canberra
Botany Bay
VICTORIA
Goulburn R.
Murray R.
AUSTRALIAN CAPITAL TERRITORY
Melbourne
INDIAN OCEAN
Bass Strait
TASMAN SEA
TASMANIA
Derwent River
Hobart
NEW ZEALAND

As Europeans, Asians, and North Americans arrived in large numbers, the population took on greater ethnic diversity. In the 1970s, Australians also became more responsive to the plight of the Aborigines, and Aboriginal land rights became a powerful political cause.

Australia, like many other nations, suffered an economic recession in the 1980s. Australia's people were facing high unemployment rates and job shortages, which prompted the government to limit immigration. But in the early 1990s, the government took measures to improve the standard of living for all Australian people. The government continues its efforts to maintain a strong economy and domestic security. Meanwhile, Australians renew their commitment to a healthy, diverse society through increased education and social and cultural programs.

THE LAND

Australia is the only nation in the world to occupy a whole continent. The country, which lies entirely within the Southern Hemisphere, is 2,000 miles (3,200 kilometers) southeast of mainland Asia. Australia's territory of almost 3 million square miles (7,690,000 square kilometers) is about the same size as the mainland United States. In area, Australia ranks as the sixth largest country in the world.

At its greatest distances, Australia stretches 1,950 miles (3,140 km) north to south and 2,475 miles (4,000 km) west to east. The mainland contains five of Australia's six states: Queensland, New South Wales, Victoria, South Australia, and Western Australia. The continent also includes the Northern Territory, a largely barren region that occupies almost one-fifth of the land area. Tasmania, an island southeast of the mainland, is the sixth Australian state.

With inlets included, Australia's shoreline is about 22,800 miles (36,700 km) long. North of the mainland, from west to east, are the Timor, the Arafura, and the Coral Seas. The South Pacific Ocean and the

Tasman Sea lie to the east, and the Indian Ocean touches the country's southern and western shores. Bass Strait separates Tasmania from the mainland, leading westward to the Great Australian Bight—a 600-mile-wide (966-km) bay on the continent's southern coast. The 400-mile-wide (644-km) Gulf of Carpentaria indents the continent's northern edge.

Topography

Australia has three distinct geographic regions: the Great Western Plateau, the Central Lowlands, and the Eastern Highlands (also called the Great Dividing Range). The Great Western Plateau occupies the western three-fifths of Australia. Most of the plateau is outback—the term Australians use for the dry, sparsely inhabited interior parts of the continent. The Eastern Highlands form a band down Australia's eastern coast. Separating the western and eastern regions is the Central Lowlands, which lies above a vast reserve of underground water called the Great Artesian Basin.

Within the Great Western Plateau is the entire state of Western Australia, much of the Northern Territory, a great part of South Australia, and a portion of western Queensland. Very flat overall, the region has an average elevation of 1,000 feet (305 meters) above sea level. In its center are three deserts, the Great Sandy, the Gibson, and the Great Victoria. The Great Sandy and the Great Victoria Deserts are areas of swirling sands and giant dunes. To the north and southwest of the deserts are hilly scrublands that can support livestock.

Australia is very old geologically. Some rocks on the Great Western Plateau are more than 3 billion years old. This means the rocks have been around since shortly after the earth's surface first cooled and solidified.

The wettest and most fertile sections of the Great Western Plateau are the coastal plains in the far north and in the southwest. A 400-mile-long (644-km) treeless plateau called the Nullarbor Plain extends along the southern edge of the plateau north of the Great Australian Bight.

The Central Lowlands cover much of Queensland and New South Wales and parts of Victoria and South Australia. Except for the coastal sections in the north and south, the lowlands are too dry and hot for crop farming. Inland riverbeds are empty most of the year. The region supports a sheep industry, however, due to the Great Artesian Basin, from which wells draw water for livestock. The barren Simpson Desert forms the west central part of the lowlands. At the southern edge of this desert is Lake Eyre, which at 52 feet (16 m) below sea level is the lowest point in Australia.

The dry and barren **Central Lowlands** support very little plant life.

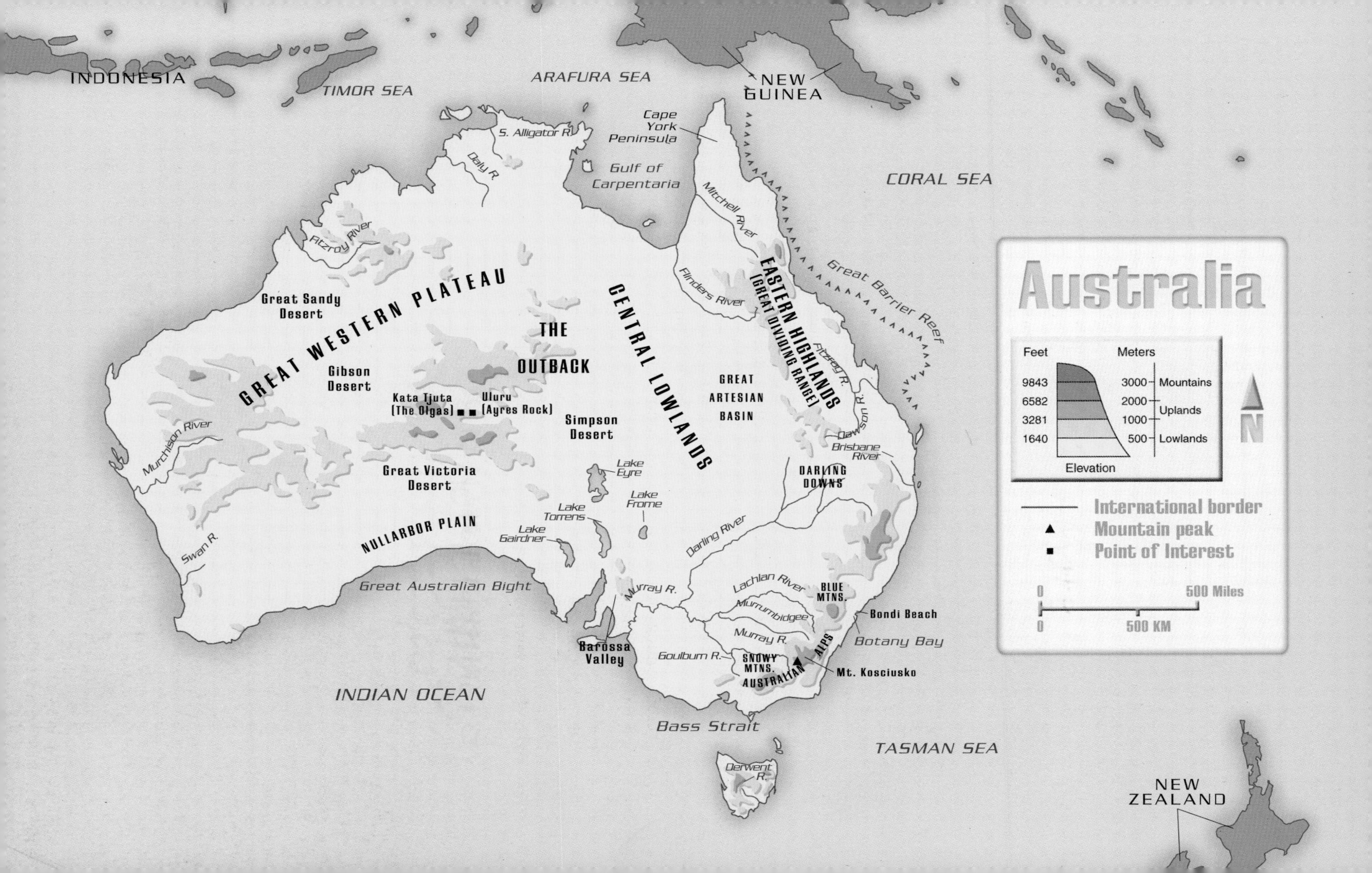
Australia
Feet
9843
6582
3281
1640
Meters
3000 Mountains
2000
Uplands
1000
500 Lowlands
Elevation
International border
Mountain peak
Point of Interest
0
500 Miles
0
500 KM
N
INDONESIA
TIMOR SEA
ARAFURA SEA
NEW GUINEA
CORAL SEA
Cape York Peninsula
Gulf of Carpentaria
S. Alligator R.
Daly R.
Fitzroy River
Mitchell River
Flinders River
Great Barrier Reef
EASTERN HIGHLANDS
(GREAT DIVIDING RANGE)
Fitzroy R.
Dawson R.
Brisbane River
GREAT WESTERN PLATEAU
Great Sandy Desert
Gibson Desert
THE OUTBACK
CENTRAL LOWLANDS
GREAT ARTESIAN BASIN
Kata Tjuta (The Olgas)
Uluru (Ayres Rock)
Simpson Desert
Murchison River
Great Victoria Desert
Lake Eyre
Lake Frome
Lake Torrens
Lake Gairdner
DARLING DOWNS
Darling River
NULLARBOR PLAIN
Swan R.
Great Australian Bight
Murray R.
Lachlan River
BLUE MTNS.
Murrumbidgee
Murray R.
Bondi Beach
Botany Bay
ALPS
Barossa Valley
Goulburn R.
SNOWY MTNS.
AUSTRALIAN
Mt. Kosciusko
INDIAN OCEAN
Bass Strait
TASMAN SEA
Derwent R.
NEW ZEALAND

The Eastern Highlands extend from Cape York Peninsula in the northeast to the southern coast of Tasmania. The coastal areas of Queensland and New South Wales, much of Victoria, and the island of Tasmania lie in the Eastern Highlands. This region of low mountains, plateaus, and coastal plains contains some fertile soil and receives more rain than the rest of the country. Most Australians live in the southern half of this area.

Australia's east coast also supports the continent's rain forests. These pockets of dense vegetation and rare birds and animals are believed to be survivors of an ancient rain forest that once covered most of Australia. The chain of rain forests begins in northern Queensland, on the Cape York Peninsula. Queensland's rain forests are tropical—hot and humid, with high rainfall concentrated in a four-month period. As the forests continue down the east coast to Tasmania, they change to subtropical and temperate zones, with less rain and more moderate temperatures.

Rain forest

The highest elevations in Australia are in the far southern part of the Eastern Highlands. Mount Kosciusko—the country's tallest peak at 7,310 feet (2,228 m) above sea level—rises from the Australian Alps in the southeastern corner of the continent. Across Bass Strait, mountains also dominate the island of Tasmania. This state contains several peaks that exceed 4,000 feet (1,219 m) above sea level.

The Great Barrier Reef, the largest deposit of coral in the world, stretches for 1,250 miles (2,012 km) along the northeastern coast of Australia. Coral is a sea creature with a hard outer skeleton. Skeletons of dead coral will build up in warm shallow seas, forming reefs, underwater walls that are home to many rare forms of sea life. The Great Barrier Reef is actually a chain of 2,600 coral reefs and 320 coral islands. In the north, the reef lies close to the mainland. It gradually spreads outward, and the southern tip is 150 miles (241 km) from the continent.

The enormous coral deposits of **the Great Barrier Reef** can be seen even from above.

Flowing through southeastern Australia, the Murray River forms the border between Victoria and New South Wales.

Water Resources

Major underground basins that collect and store water lie beneath 60 percent of Australia's land surface. People and livestock in interior regions rely on wells that tap these stores of groundwater. The greatest reserves are in the Great Artesian Basin. Much of the water in this basin is salty, however, and can be drunk only by livestock.

The 1,600-mile-long (2,575-km) Murray River and its tributaries—the Darling, the Murrumbidgee, the Lachlan, and the Goulburn—form the country's largest river system. This network of waterways extends over a large part of southeastern Australia. These rivers supply 80 percent of Australia's irrigated land with water.

The Snowy Mountains Scheme, completed in 1974, is a massive water-conservation and hydroelectric project. Aqueducts (canals) and tunnels collect the flow from melting snows in the Snowy Mountains, a range in the Australian Alps. These channels carry the runoff to dams and reservoirs for use during dry periods. Other tunnels redirect mountain streams that formerly ran eastward into the Tasman Sea. Rerouted to the Murray and Murrumbidgee Rivers, this additional freshwater increases the ability of these rivers to irrigate southeastern Australia.

Lake Eyre, situated in the outback of South Australia, is the country's largest lake. Because of infrequent rains, however, this desert

The dry, cracked floor of South Australia's salty **Lake Eyre** is exposed for much of the year.

lake's floor—called a playa—is usually dry. Floods have allowed the huge salt lake to reach its full size only twice since Europeans began keeping records about the surrounding desert. Other large playas in South Australia form Lake Torrens, Lake Gairdner, and Lake Frome.

Climate

A hot, tropical climate marks most of northern Australia. The southern half of the continent is in the temperate zone. Because Australia lies in the Southern Hemisphere, summer occurs from December to February, and the winter months are June, July, and August.

In the north, the average temperature is 84°F (29°C) in January (the hottest month for most of the country) and 77°F (25°C) in July. In the south, those same months bring average readings of 63°F (17°C) and 46°F (8°C). Some parts of the interior of Australia, however, see average daily highs above 100°F (38°C) in January. When summer winds blow from the interior, the cities on the eastern coast get very warm.

Winters can be chilly in the south, and mountain ranges there receive large amounts of snow. The coldest regions of Australia are in the highlands of Tasmania and in the southeastern corner of the continent. In central Australia, winter brings cool nights, but the days are usually warm and dry.

Australia receives the least rainfall of any inhabited continent. Half of Australia sees fewer than 11 inches (28 centimeters) of rainfall annually. Some areas may not have rain for years. Most of the country's riverbeds are dry for part of the year, filling only during the rainy season. In northern Australia, the wet period occurs during the summer, from December to February. In the south, rain falls primarily in the winter, from June to August. Moisture evaporates rapidly throughout much of the continent, but floods are common in important agricultural areas.

Flora and Fauna

Because it is isolated from other large landmasses, Australia has unique plants and animals. In fact, 80 percent of Australia's flora and fauna are found nowhere else in the world. Environmental factors, such as low rainfall and sparse food supplies, have also produced unusual features.

Australia's best-known trees are the gum (also called eucalyptus) and the wattle (acacia). They grow in all regions of the country except the driest deserts. Of the 550 species of gum trees, the jarrah is prized for its extremely hard and durable wood. More than 600 types of wattle—a flowering tree—can be found throughout the country. The yellow wattle appears on the Australian coat of arms.

Snappy gum tree

More than half of Australia's 230 species of native mammals are marsupials, or animals that develop their young in a pouch. This group includes koalas, Tasmanian devils, possums, wombats, bandicoots, and kangaroos. The tree-dwelling koala is a good example of adaptation to environmental factors. The koala feeds only on eucalyptus leaves, which are poisonous and low in protein. But eucalyptus is plentiful in Australia, and it has a high oil content, enabling the koala to go without water for long periods of time. So the koala has adapted. Its stomach detoxifies the eucalyptus leaves, and its metabolism has slowed to account for the lack of protein.

Australia's best-known marsupial is the kangaroo. The kangaroo family includes more than 50 kinds of animals. They live in every type of Australian habitat, from rain forests to rocky cliffs. The largest and most common kangaroos are the red and gray varieties, which graze in open grasslands. An adult male can stand over 6 feet (1.8 m) tall and weigh 180 pounds (82 kilograms).

Two **gray kangaroos** take a morning hop through their home in South Australia. This type of kangaroo has a long-haired and silver gray coat in the eastern coastal regions, but a short-haired and dark gray coat inland.

The **short-beaked echidna** is one of only three egg-laying mammals on earth. Visit vgsbooks.com to learn more about Australia's fascinating wildlife.

Most mammals give birth to and nurse live babies. But some mammals native to Australia are unusual because they lay eggs. Monotremes, as these species are called, lay eggs but then nurse after the eggs hatch. Monotremes include the echidna (spiny anteater) and the duck-billed platypus.

Australia also has familiar nonnative mammals. The dingo, or wild dog, is one of the country's most powerful predators. The first dingoes probably came to the continent from Asia with Aborigines thousands of years ago. Rabbits have also become common in Australia and are sometimes so numerous that they destroy precious pastureland. Despite efforts to control their population, rabbits flourish.

At least 700 species of birds are native to Australia. Among them are black swans, flightless giant emus, and about 60 kinds of parrots—including cockatoos and parakeets. Many Australian birds are noted for unusual songs and behaviors. The lyrebird can mimic the songs of other creatures. The male bowerbird builds bowers, or arched shelters, in which it performs elaborate courtship dances to attract females. The kookaburra's call sounds like human laughter ringing through the forests.

Kookaburras are best known for their laughter-like call. To let other kookaburras know about their home territory, a family group will laugh throughout the day.

Natural Resources

Australia contains rich deposits of many metals and minerals. Its supply of iron ore ranks among the largest in the world. The country is also a major producer of bauxite, the principal source of aluminum. Great reserves of coal and uranium exist in Australia, and supplies of natural gas and crude oil have been developed.

Australia is a major supplier of lead, zinc, copper, and nickel, as well as an important source of mineral sands used to make heavy metals. The country also contains tin, silver, and some precious gems, including diamonds and opals. Gold—the metal that touched off massive immigration to Australia in 1851—is still mined in Western Australia and Queensland.

Major Urban Centers

Despite its large size and low population density, Australia is a very urbanized country. Eighty-five percent of its 19.2 million people live in cities and suburbs. Five metropolitan areas have populations exceeding 1 million. Four are situated in the southeastern part of the country. The fifth is Perth, a coastal city in the southwestern corner of Australia.

SYDNEY New South Wales's capital is home to 20 percent of the country's population. Established in 1788 as a colony for British prisoners, the settlement developed into a major seaport after farming products began to

Modern office buildings dominate the Sydney skyline.

be exported. The long port, which has many bays, reaches inland for 21 miles (34 km), dividing the city into northern and southern sections. The Sydney Harbour Bridge, which opened in 1932, connects the two halves. The world-famous Sydney Opera House sits on Bennelong Point, at the southern edge of the harbor. Sydney is also Australia's largest manufacturing center.

MELBOURNE is the nation's second largest city, with 3.4 million people. It is the capital of Victoria. Founded in 1835, Melbourne grew rapidly after gold was discovered in the state in the 1850s. This business and industrial hub offers many cultural opportunities. Melbourne and its suburbs have several universities, fine museums, and medical and scientific research institutions.

BRISBANE With a population of 1.6 million, Brisbane sits in the southeastern corner of Queensland and is that state's capital. Great Britain founded Brisbane as a colony for convicts in 1823 and maintained it as a prison site until 1839. The British opened the town to free settlement in 1842. A warm climate helps the modern city attract tourists from southern Australia in winter.

PERTH is the capital of Western Australia. With 1.4 million residents, the nation's fourth largest city lies on the Indian Ocean near the southwestern corner of the continent. Fifteen hundred miles (2,400 km) of outback separate Perth from Adelaide, the closest large city. Perth has grown rapidly in recent years because of an increase in commercial activity and tourism. Western Australia's reserves of iron ore, oil, natural gas, bauxite,

Characterized by numerous waterways and a compact city center, Perth is also the home of the world's oldest operating mint.

and other minerals are boosting Perth's economy and population.

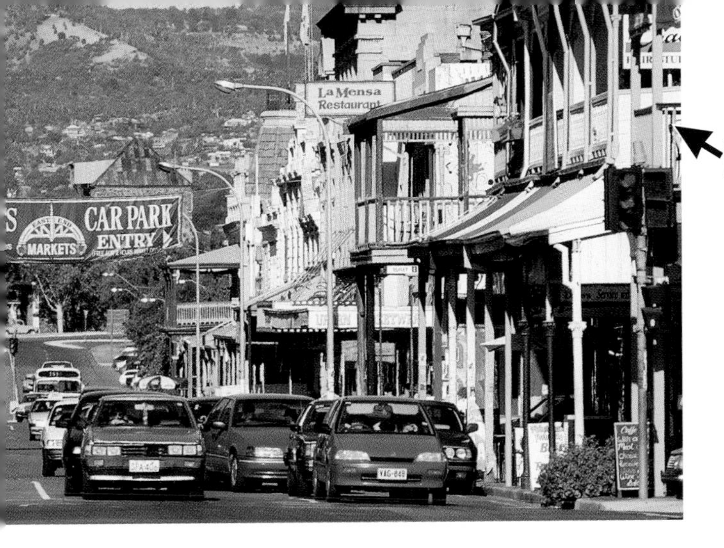

Adelaide, the capital of South Australia, is home to 73 percent of the state's total population.

ADELAIDE The capital of South Australia is sheltered by the Mount Lofty Ranges. This city of 1.1 million people has been an agricultural- processing center for much of its history, but the manufacturing of products such as cars and appliances has become the chief economic activity. Tourism is also important to the city's economy. Nearby Port Adelaide is a center for submarine manufacture.

SECONDARY CITIES Canberra is the capital of Australia. It is a planned city, specifically built to house the federal government and centered on Parliament House. Building began in 1913, after architects from all over the world submitted plans. Situated in southeastern New South Wales, the city has 309,300 residents. In addition to governmental offices, Canberra contains many research institutes, the Australian National University, and a military college. The capital's land area forms the Australian Capital Territory (ACT), a political unit administered by the national government, separate from New South Wales. The nearby Territory of Jervis Bay is a strip of land set aside in 1915 that gives the ACT access to the sea.

Hobart, the capital of Tasmania, was founded in 1803 as a prison colony for lawbreakers who were repeat offenders. By the mid-twentieth century, Hobart had become a major whaling port and boat-building site. With a population of 194,200, the city serves as a center for both agriculture and industry.

Two important towns in the Northern Territory are Darwin, the region's capital, and Alice Springs. With a population of 88,100, Darwin sits at the northern edge of the outback, on an inlet of the Timor Sea. Darwin is a key defense base for Australia, and military personnel make up a large part of the population. Mining and tourism in the Northern Territory have boosted Darwin's economic importance in recent years. Alice Springs, at the southern end of the territory, is close to the geographical center of the continent. This town of about 27,000 residents is a center for the outback cattle industry. Also a tourist destination, Alice Springs is a departure point for trips to Uluru (Ayers Rock), the world's biggest rock.

HISTORY AND GOVERNMENT

At least forty thousand years ago, Stone Age peoples reached Australia from Southeast Asia. The migrations to Australia came in waves over hundreds of years. Centuries later, Europeans were to call these groups Aborigines (from the Latin phrase *ab origine*, meaning "from the beginning").

The earliest migrants probably crossed over land and water from the Asian mainland. Eventually, however, migrations ceased, and the Aborigines lived in isolation from other peoples for thousands of years.

The First Australians

About five hundred Aboriginal groups, each with a distinct language, established themselves in Australia. They lived by hunting and gathering. Some families built shelters of branches and grasses, but they had no permanent dwellings. Numbering about 300,000, the Aborigines lived primarily along the northern and eastern coasts, in the Murray River Valley, and on the island of Tasmania.

Each Aboriginal group occupied a recognized territory. Within the boundaries of each territory, clans of ten to fifty people roamed in search of food. The Aborigines obeyed strict family rules and customs, with the young and strong providing for the old and feeble. Men of high standing in each group led ceremonies and made decisions affecting their people.

Among the Aborigines' few possessions were digging sticks, bags for carrying food, and didgeridoos—musical instruments that produce a wailing sound. Their hunting weapons included boomerangs, traps, nets, barbed wooden spears, stone axes, and pointed sticks.

Following strict rituals that inflicted little harm, Aborigines used warfare to show their skill and daring. Meeting at traditional battle sites, warring groups would exchange insults and challenges and then hurl spears at the opposing side. By confronting the enemy and dodging weapons, each side achieved prestige.

European Arrival

Until the seventeenth century, foreign influences had little impact on Aboriginal culture. However, expanding trade in Asia led European powers—particularly the Netherlands—to take a greater interest in the South Pacific. Many navigators believed in the legend of a "great southland" full of precious metals and exotic spices.

In pursuit of this legendary continent, the Dutch navigator Willem Jansz reached northeastern Australia in 1606. Jansz thought the area was an unknown part of New Guinea, a large island north of Australia. Although his reports about the region and its people were discouraging, Dutch explorers continued to search for the great southland. In doing so, they charted much of the western coast of Australia. In 1642 Abel Tasman sailed around the continent to Tasmania, which he named Van Diemen's Land. Tasman shared Jansz's unfavorable impressions of the area.

Abel Tasman

The Dutch explorations encouraged visits by other European navigators. The first British explorer to arrive in Australia was William Dampier, in 1688 and again in 1699. He, too, found the territory unpromising and doubted that it could be the great southland of legend. Interest in the continent lagged until 1770. In that year, Captain James Cook, exploring for the British navy, charted the east coast down to the southeastern corner.

On April 19, 1770, Cook's ship, the *Endeavour*, reached Botany Bay, which Cook named for the many unusual plants growing there. The navigator claimed the eastern part of the continent for Britain. Exploring the South Pacific again from 1772 to 1775, Cook proved that Australia must be the great southland of legend.

British Settlement

At first the British made no attempt to colonize Australia. In 1783, however, the United States won its independence, and Britain lost its American colonies. Until this time, Britain had used two of those colonies—Georgia and Maryland—as prisons for its convicts. Britain had few prisons itself and needed a new place to send people convicted of crimes. Officials decided the remote continent of Australia would suit the purpose.

After an eight-month journey from Britain, the first fleet of eleven ships arrived in Botany Bay in January 1788. The vessels carried 548 male

Convicts on a ship bound for Botany Bay. During the late 1700s (and for decades to follow), the British shipped criminals to Australia for punishment.

and 188 female convicts and about 300 free persons, most of whom were soldiers serving as guards. Captain Arthur Phillip, the colony's governor, decided to move the fleet seven miles (11 km) north, where he hoped to find a better harbor and richer soil. On January 26, 1788, the settlers arrived at what would become Sydney, New South Wales, and founded the first white settlement on the continent.

From the beginning, the colony at Sydney faced great difficulties. Phillip had expected that prisoners would soon be growing food. But the soil in the immediate area was poor. Wheat seeds had been damaged on the journey and failed to grow. Animals brought on the first fleet died or escaped.

The practice of shipping convicts to British colonies was called transportation. Until 1783 the prisoners were sent to the American colonies. After the American Revolution, they were sent to Australia. So many criminals were sentenced to transportation before the practice ended that both Americans and Australians were often considered "a race of convicts."

Australia's largest and most famous city, **Sydney,** started as a British prison colony. Convicts and their guards lived in hastily built barracks.

Convicts and their keepers alike were near starvation when a second fleet finally arrived two and one-half years after the first. The second fleet, however, had little food to spare, and many of its passengers were ill or dying.

To encourage farming, the governor began giving land to soldiers and to those convicts who had finished serving their terms. To obtain enough land, the officials seized territory from local Aborigines, offering no payments or treaties in return.

It was not easy to maintain order in this new colony. A special military force called the New South Wales Corps was brought in from Britain. Officers and soldiers in the corps treated the convicts cruelly. Conditions were especially harsh on Norfolk Island, an outpost situated 1,000 miles (1,609 km) east of Australia that had been set up to hold second offenders and political prisoners. Some inmates there committed murder so that they would be hanged and escape their misery.

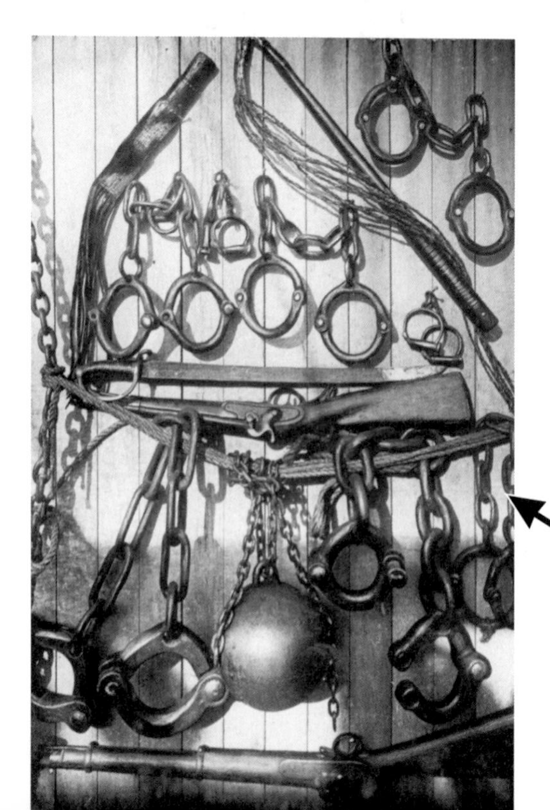

Shackles, whips, and clubs were among the implements used to discipline the convicts housed on Norfolk Island.

Early Reforms

In 1792 Captain Phillip returned to Britain, and officers of the New South Wales Corps gained power in the colony. The officers used their authority to enrich themselves. They acquired land and gained control of the rum trade. Rum originally arrived on trading ships from America. The officers bought the cargo, then sold it to colonists at inflated prices. The rum eventually became the colony's currency. Farmers were forced to accept the liquor in exchange for their crops, and in turn they had to pay their workers in rum. The officers of the New South Wales Corps, who profited most from this system, came to be known as the Rum Corps. While the Rum Corps controlled affairs, little consideration was given to the rights and needs of freed convicts.

Reforms in Australia took hold, however, after Lachlan Macquarie became governor in 1810. Macquarie broke the Rum Corps's monopoly and established a banking system. He constructed public buildings and roads. Macquarie supported the rights of former convicts, many of whom became productive citizens after serving their prison terms.

In 1813 Macquarie also encouraged the first crossing of the Blue Mountains. A rugged chain in the Great Dividing Range, the Blue Mountains had confined settlement to a semicircle of land around Sydney. With a route open through the mountains, free settlers and colonial officers began occupying grasslands farther west.

The colony's first military governor, Scottish-born **Lachlan Macquarie** helped destroy the Rum Corps's monopoly and encouraged exploration of Australia's uncharted territories. Visit vgsbooks.com to find out about a popular tourist spot in Sydney, Mrs. Macquarie's Chair, named for the governor's wife.

Although the government claimed ownership of the land, the people moving in asserted their right to it by "squatting"—staking claim by living on the property. Some squatters seized huge tracts—as much as 20,000 acres (8,000 hectares). Frequently, they obtained land by killing the Aborigines who lived in the area or by using guns to drive them away.

Such actions by squatters added to the problems of the Aboriginal population, which began to shrink soon after Europeans settled in Australia. Many Aborigines died from smallpox and other diseases unknown to them before the British arrived. Liquor—easy to obtain in the colony—ruined the lives of many other Aborigines. The newcomers competed with the Aborigines for the dwindling supply of fish, and they drove away kangaroos and other game that the Aborigines hunted for food.

While the Aboriginal population disappeared from southeastern Australia, the region's squatters prospered. Most landholders began raising merino sheep, a breed that could survive on dry vegetation. These sheep produced a fine wool that settlers could sell abroad.

Exploration and Expansion

As the settlement grew, explorers journeyed north of Sydney, discovering the Darling Downs area of southern Queensland and the Brisbane River.

Squatters and farmers **raised merino sheep for wool** in the early 1800s. Merino wool continues to be a staple export for Australia.

Two adventurers walked overland to the southern coast, arriving at what would become Melbourne. To prevent other European countries from settling Australia, Britain claimed the western half of the continent in 1829.

Setting out the same year, Captain Charles Sturt traced the course of the westward-flowing rivers and learned that Australia had no great inland sea, as some had thought. This was the first of many expeditions that attempted to unlock the secrets of the continent's interior. Some explorers who challenged the outback became folk heroes. Others perished in their efforts to cross the hot, parched land.

The dry climate of inland Australia limited most settlement to the edges of the continent, where Australia's first cities were established. After Sydney, Britain established the prison colony of Brisbane in 1824 on the Brisbane River. A settlement that sprang up in 1829 on the Swan River in Western Australia eventually became the city of Perth. Sheep farmers from Tasmania founded Melbourne on the mainland in 1835. These squatters did well, and Britain later granted their request to form Victoria, a colony separate from New South Wales. The South Australia Company in London established the colony of South Australia, with its capital at Adelaide, in 1836. In addition to the British, South Australia's earliest settlers included Germans who were seeking freedom from religious persecution in their homeland.

Down Under

The name Australia comes from *terra australis*, Latin for "southern land." The phrase was first used by a British navigator, Matthew Flinders, in the late eighteenth century. Flinders sailed all the way around the continent surveying its coasts. The map he published as a result of that journey was the first document on which the name Australia appeared. The Southern Cross constellation has also become a symbol for Australia, appearing on its flag. The constellation guided sailors through the seas around Australia. The country became known as the Land Down Under because it is in the Southern Hemisphere, on the other side of the world from Great Britain.

The Gold Rush

Sheep raising and the export of wool to Britain dominated Australia's economy in the first half of the nineteenth century. In 1850 Australia had 18 million sheep and only 400,000 people. In 1851, however, Edward Hargreaves, a mining prospector, discovered gold at Bathurst in New South Wales. By the end of the year, prospectors found even bigger deposits in Victoria.

Until these mining discoveries, Australia had difficulty attracting enough immigrants to meet its labor needs. But the possibility of finding gold suddenly brought prospectors from Europe, the United States, and China across great distances to the South Pacific.

Just as Australia was growing with the gold rush, the British Parliament passed legislation in 1855 allowing the colonies to become largely self-governing. They were still under British authority, but administering the distant colonies was difficult and expensive for Great Britain. By 1859 the four eastern colonies—New South Wales, Victoria, South Australia, and Tasmania—had adopted constitutions and formed their own governments and legislatures. Queensland separated from New South Wales in 1859 and also became self-governing. (Western Australia did not adopt limited self-government until 1890.)

The gold rush helped to promote a sense of social equality among white Australians. Prospectors could become wealthy by striking gold and did not have to depend on family connections to succeed. In fact, people accustomed to hard physical work had an advantage in gold prospecting over those not used to manual labor. The living conditions

The gold rush of the 1850s gave tough, strong laborers a chance to raise their fortunes and social status by striking it rich.

in the goldfields were rugged for everyone and blurred class barriers. Throughout Australia, as members of the working class grew wealthier, they began to fill seats in the colonial parliaments.

But the easing of social barriers did not benefit everyone during the gold rush. For many, there was still an unquestioned belief that the country should be settled and developed by white people of European descent. Asians who came to Australia to work the goldfields were met with open hostility.

The town of Kalgoorie in Western Australia is the country's largest producer of gold. The Golden Mile vein, one of the richest gold lodes ever found, was discovered near Kalgoorie in 1883. More than a century later, most of the town's 20,000 residents still earn their living from mining the vein.

Despite these conflicts, the gold rush changed Australia. By drawing people of widely different occupations, the discovery of gold paved the way for industrialization. It also helped boost the population past one million by 1860. With a more abundant labor supply, large landholders no longer needed British convicts. The shipping of convicts to Australia—a practice that had brought 160,000 people to the colony—ended in 1868.

Forming One Nation

Soon after achieving limited self-rule, the colonies began holding regular conferences that gave the prime ministers a chance to discuss common concerns, such as taxing imports. Considerable rivalries existed among the colonies, however, and close cooperation and political union was slow to develop.

In the early 1890s, the colonies experienced an economic depression. As people lost their jobs, membership in trade unions increased. The unions formed political parties in each colony to push for laws that would help working people. These labor parties wanted the colonies to unite because they believed that standardized labor laws would benefit workers.

Meanwhile, the colonies saw the need to set up common immigration laws and tariffs (import taxes). They decided to form a federation, giving most powers to the national government but reserving some authority for the states. In 1897 delegates to a national convention began to draw up a constitution, which voters in all colonies had accepted by 1899. Britain approved the plan in 1900, and the Commonwealth of Australia was formally established on January 1, 1901. Melbourne served as the capital temporarily, while the government built the permanent capital of Canberra.

AUSTRALASIAN FEDERAL REFERENDUM
JULY 1899

DIEU ET MON DROIT

This is to Certify that the question of an Australasian Federal Constitution was submitted to the Electors of Victoria on the 27th day of July 1899 and that out of a Poll of .... votes .... WERE CAST IN FAVOR OF SUCH AUSTRALASIAN FEDERAL CONSTITUTION and further that Mr. ________ VOTED AT THE SAID ELECTION.

A copy of this **voter's certificate** was given to every man who voted in favor of accepting Australia's new constitution in 1899.

Although the new federation was an independent nation, rivalries among the former colonies continued after they became the six states of Australia. Often, the states trusted Britain more than they trusted one another, so even after independence they chose to maintain political ties to Britain. As a member of the British Commonwealth, Australia continued to recognize Britain's monarch as its own head of state. Britain still exercised authority over Australia's foreign policy, and British laws could overrule legislation enacted by the Australian Parliament.

In 1901 the new government adopted immigration laws virtually excluding non-Europeans from Australia for more than fifty years. Unofficially known as the White Australia policy, the legislation reflected a strain of racial prejudice among Australians that first surfaced during the gold rush.

When Australia was established, about 60,000 Aborigines were living in the country. The new federal and state governments adopted policies to protect Aborigines but did not give them citizenship rights. Most lived on reserves (land set aside) in the barren Northern Territory and in Queensland, away from white Australians. In general, the government ignored the needs of these people. Pushed from their traditional way of life, the Aborigines came to depend on churches and other private groups for assistance.

The World Wars

Australia's political ties to Britain had an important effect in 1914, when World War I began. When Britain declared war against Germany that year, Australia was at war as well. The country sent more than 330,000 soldiers to fight in Europe and northern Africa. Australians and New Zealanders fought jointly as the Anzac forces (Australians and New Zealand Army Corps). The war cost the lives of nearly 60,000 Australians. Yet their achievements in the war helped Australians develop a sense of common purpose and nationhood.

World War I strengthened Australia economically as well. The demand for wartime goods encouraged manufacturing and boosted exports of food, minerals, and wool. European immigration to Australia resumed after the war ended in 1918, and the 1920s brought improvements in the standard of living. The nation built more roads to accommodate a growing number of cars. The cities acquired electric trains. Telegraph lines, telephones, and airmail letter service put Australians in closer touch with other parts of the world.

The Great Depression of the 1930s, however, ended this upswing. As the world's economy sank, many large Australian companies went bankrupt. In some cities, 30 percent of the people lost their jobs, and immigration virtually ceased. The population grew so sparse that many Australians feared the country had too few people to survive.

Anzac forces bring supplies ashore at Gallipoli, Turkey, where Australians fought a heroic, though doomed, battle during World War I.

When World War II broke out in 1939, Australians again joined Britain as an ally. Allied forces were soon fighting not only in Europe against Germany and Italy but also in the Pacific against Japan. The threat that Japan would invade Australia was very real. After the British colony of Singapore fell to the Japanese in February 1942, Australians realized that British forces in the Pacific could not protect them from invasion. Australia turned to the United States for help.

John Curtin, the Australian prime minister, placed Australia's forces under the command of U.S. general Douglas MacArthur. One million U.S. soldiers moved into the country, whose own population was only seven million. Japanese planes bombed the Australian cities of Darwin and Broome in early 1942, and Japan seemed to be winning the war in the Pacific. But in May 1942, a combined U.S.-Australian fleet won an important victory against Japan in the Battle of the Coral Sea, and the tide began to turn. Still, fighting continued for three more years until Japan surrendered on August 14, 1945.

To learn more about Australia's history, including Captain Cook, the convicts of the First Fleet, the displacement of the Aborigines, the gold rush and more, go to vgsbooks.com.

Postwar Growth and Prosperity

After the war, Australia and the United States strengthened the close relationship they had developed. With New Zealand as a third partner, they signed the ANZUS (Australia, New Zealand, United States) pact for mutual defense. To further protect its interests, Australia also became a founding member of the United Nations in 1949.

A long period of political stability began in 1949 when Robert Menzies became prime minister. Menzies was the leader of the Liberal Party, which had formed during World War II to oppose some of the Labor Party's policies. Although he was sometimes viewed with more awe than affection, Menzies led the country for a record sixteen years, helping to shape Australia's postwar progress.

During this period of prosperity and development, Menzies was often seen as a defender of traditional Australian values. His party concentrated on the needs of the middle class, as the "backbone of the nation." But he also took an active interest in Pacific and Southeast Asian affairs and sought to prevent Australia's isolation in the region.

Under his leadership, immigration policies began to change. After the war, concerns arose about both military defense and severe labor shortages. Australians again feared that there simply were not enough people in the country to defend it or to support increasing manufacturing demands. Menzies declared that Australia must "populate or perish." To encourage immigration, the government offered incentives—such as free land in Australia and assistance with transportation expenses to get there—to people in other countries. More than 800,000 non-British Europeans, including war refugees from eastern Europe, came to live in Australia. They brought a great variety of foods, customs, and celebrations, positively influencing the idea of a diverse Australia. Not long after the influx of these groups, the White Australia policy was abandoned. Since then the country has welcomed thousands of refugees and others from Southeast Asia, the Middle East, Central and South America, and Africa.

In the 1960s, the rights of Aborigines also began to slowly move forward. They finally gained full citizenship, including the right to vote. Australians elected to give the national government, rather than the states, responsibility for Aboriginal affairs. The government began to involve Aborigines in decisions affecting their land, housing, health, education, and employment.

Decades of Change

Through most of the postwar decades, politics in Australia largely involved the Labor, Liberal, and National Parties. From 1949 to 1972, the more conservative Liberal and National Parties controlled Parliament. The standard of living greatly improved. Demand was high for Australian raw materials, and this fueled the development of mining in Western Australia. With a focus on economics and middle-class prosperity, the Liberal-National coalition remained popular.

In the early 1970s, however, world demand for the output of Australia's mines dropped, hurting the country's economy. Labor strikes added to the nation's financial problems. In some cases, workers won pay hikes that were not matched by increases in productivity (output). As labor costs increased, manufacturers had to raise prices on Australian-made goods.

Political and social issues further disrupted Australia in the 1970s. Groups demonstrated against the government's support for U.S. policies in Southeast Asia, especially the Vietnam War. Members of the women's movement sought greater participation for women in Australian political and business life. Other activists attempted to raise public concern about the social problems and land rights of the Aborigines.

In 1972, amid social protests and economic problems, national elections brought the Labor Party to power for the first time in more than twenty-five years. Led by Gough Whitlam, the Labor government started many progressive social programs. But the prime minister enacted these changes at a time when the country could not afford to pay for them. Unemployment was high, and foreign trade earnings were dropping. Many members of Parliament believed Whitlam's programs were too expensive. Eventually, Parliament refused to pass bills that required government funding.

Gough Whitlam

Usually, if a parliamentary government cannot pass legislation, the ruling party agrees to give up power, and an election is held. In this case, the Whitlam government held out, hoping for cooperation. But after a month's deadlock, Australia's governor-general, Sir John Kerr, took an unprecedented step. Using his formal authority as the representative of the British monarch, he dismissed the Whitlam government in November 1975 and installed Malcolm Fraser as a caretaker prime minister. Many Australians believed the governor-general did not have the real power to do this, and the dismissal became the most serious constitutional crisis in Australian history.

General elections were held the following month. Despite the controversy, Fraser, the head of the Liberal Party, was voted in as prime minister. Fraser's government struggled to fix the problem of inflation, but unemployment increased. By the early 1980s, after the collapse of the mineral boom and a severe drought, the country faced more inflation and unemployment. In new elections, the Labor Party regained power in 1983, with Robert Hawke as prime minister.

To improve the economy, Hawke developed a strategy of cooperation, asking groups to work together. He held a meeting of business, union, and government leaders in Canberra. The three groups signed an accord to cooperate for the good of the whole society. The accord reduced the frequency of labor strikes, restored social programs, and spurred economic growth.

Aboriginal land rights also developed as a political issue. In 1985 ownership of a well-known area containing Uluru (Ayers Rock), a sacred site, was returned to the Aborigines. The movement to return land to the Aborigines slowed when mining companies began to develop much of the disputed territory. These firms claimed that their use of the land entitled them to retain ownership.

The land-rights issue was also overshadowed by the country's continued economic problems. The standard of living fell during the 1980s as the country coped with unemployment and inflation. But Hawke continued to have good relations with both the business sector and the trade unions until 1989. In August domestic airline pilots went on strike. As the work stoppage continued into 1990, Australia's important tourism industry lost a large amount of business, and by 1991, Australia was in recession. In the summer of 1991, another member of the Labor Party, Paul Keating, challenged Hawke's leadership and eventually took over as prime minister.

Keating's popularity waned, too, and the economy did not improve. By 1996 unemployment was still high, and wages had declined. In that year's general election, Keating was defeated by a landslide. The Liberal-National coalition took power, and John Howard became prime minister.

New Challenges

An experienced government minister, Howard appealed to the middle and working classes. His government introduced economic reforms and made plans to restructure the country's budget. Restrictions on trade and finances were relaxed. Many industries that had been run by the government were privatized (sold to private companies), among them telecommunications, railroads, and airlines. When the government also slashed spending and cut finances in public welfare areas such as social security, health, and education, Australian citizens responded with demonstrations and protests. But by 1998, the government could show decreased unemployment and inflation, and the Liberal-National coalition was reelected.

Howard has come under criticism for backing away from the campaign for reconciliation between white and Aboriginal Australians. The government has allocated several hundred million dollars to be spent on improving health care, settling legal issues, and resolving social injustices in Aboriginal communities. Yet human rights groups remain critical of Australia's record of neglect and abuse. Howard has also refused to formally apologize for the country's historic treatment of Aborigines, arguing that modern Australians cannot be blamed for the past and should not be consumed by a "national guilt." Although it is not a simple issue, Australians see ethnic diversity as

Throughout 2001, Australia celebrated its Centenary of Federation. The colonial and territorial governments had united as the Commonwealth in 1901. Tens of thousands of people attended the Centenary events and celebrations, held all over the country. Many Australians saw the Centenary not only as an opportunity to remember the country's history, but also as a chance to discuss its future.

one of their country's strengths, and they continue to work on reconciling the past while still moving forward.

Changes in nearby countries have also affected Australia. Under both Keating and Howard, Australia has tried to establish leadership and cooperation among Asian nations. Australia has helped China, Cambodia, and Vietnam break their international isolation. It has also developed strong trade and investment ties with Singapore, Japan, South Korea, and Taiwan. Australian businesses have invested heavily in Southeast Asia, and those ties account for roughly half of the country's imports and exports. At the same time, cultural differences sometimes hamper relations, and Australia's alliance with the United States is troublesome to some Asian leaders.

Another outcome of wider regional political and economic shifts is the enormous number of refugees seeking asylum in Australia. Many refugees arrive legally and are resettled by the government. However, recent years have seen an increasing number of illegal refugees. Desperate to escape starvation and violence in countries such as China, Afghanistan, and Iraq, they are smuggled into Australia without money or immigration papers. While the government tries to process these asylum claims, refugees are held at detention camps, usually for a few months. Many of the camps are isolated, and conditions are dismal. Refugee protests and escapes are increasing, some with the support of Australian activists. But the Howard government maintains that it owes its first duty to the security, health, and welfare of the country, and that it cannot encourage human trafficking. While embracing their diverse culture, a majority of Australians have expressed agreement with firm government controls on refugees and immigrants.

Australia remains in the position of reconciling its geographic location near Asia, its security and trade concerns, and its democratic policies. Just how serious that reconciling is was demonstrated on

STRAINED ALLIANCES

Internal political turmoil in nearby countries has strained relationships within Asia. Indonesia is Australia's largest neighbor, but relations have been difficult for many years. Indonesia has been accused of serious human rights violations, especially in regard to its annexation of East Timor, a neighboring island. Australia supports East Timor's independence, and some Asian countries are critical of that stance. Similarly, relations with China and North Korea are affected by Australia's concerns over nuclear proliferation and human rights violations in those countries.

Australian investigators inspect the remains of a car believed to have brought explosives to a nightclub in Kuta, Bali, where 200 people were killed. Many of those who died were Australian tourists.

October 12, 2002, when suicide bombers attacked a nightclub on the Indonesian island of Bali, a popular vacation spot for Australians. A total of eighty-six Australians died in the bombing, most of them young tourists and amateur athletes. The terrorist group al-Qaeda later claimed responsibility for the attack, citing Australia's involvement in the U.S.-led war on terrorism in the wake of the September 11, 2001, al-Qaeda attacks in the United States.

In the aftermath of the Bali bombing, John Howard stated that he would consider preemptive military strikes, if there were no other way to stop a terrorist attack on Australia. The statement outraged leaders of neighboring Asian countries and was controversial in Australia. Some Australian ministers, however, refused to criticize Howard, suggesting that the prime minister cannot play politics when the security of the country is at issue. Australia will continue to develop strategies that balance foreign relations with domestic concerns.

Government

Australia has governments at the federal, state, and local levels. The Federal Parliament handles national legislation (lawmaking) as set forth in a written constitution. The constitution also specifies which powers the federal government shares with the states. All other authority is reserved for the states. The British monarch, who is also the monarch of Australia, is represented by the Australian governor-general and by the state governors, who all perform largely symbolic functions.

Australia is divided into eight states and territories, all represented in the Federal Parliament. States on the continent include New South Wales, Victoria, Queensland, South Australia, and Western Australia. Also included are the Northern Territory and the Australian Capital Territory (ACT), the seat of the government. The island of Tasmania is the eighth state.

Each state has a capital city and government, administering its own system of education, transportation, law enforcement, health services, and agriculture. The federal government collects all taxes and gives a share to each state. States needing additional funds to provide services must apply to the national government.

The Federal Parliament consists of the House of Representatives and the Senate. It is the country's legislative, or lawmaking, body. The House has 150 members elected for three-year terms. In the Senate, each state has twelve senators, and the Australian Capital Territory and the Northern Territory each have two. Senators serve six-year terms for states and three-year terms for territories. Except on bills relating to financial matters, the Senate has the same authority as the House in legislation.

The Parliament Building is located in Canberra, Australia's national capital.

Australia's Commonwealth **Coat of Arms** is rich in symbolism.

The Commonwealth Coat of Arms

The shield in the center of the coat of arms contains the badges of the six Australian states and is a symbol of the 1901 Federation. The seven points on the gold star above the shield represent the six states and the Commonwealth territories. Supporting the shield on each side are native Australian animals, the red kangaroo and the emu. Usually the coat of arms is depicted on a background of golden wattle (acacia) branches, but that is not part of the original design. The Australian government uses the coat of arms to authenticate documents and for other official purposes.

Voters, who must be at least 18 years old, normally choose national representatives every three years. The party or coalition of parties with a majority in the House of Representatives runs the government and chooses the ministry—prime minister and cabinet—from its membership. The ministry can continue in office as long as Parliament continues to pass the ministry's bills.

The High Court of Australia heads the judicial system. A chief justice and six other justices hear appeals from lower courts and cases involving the interpretation of the constitution. Australia has both state and federal court systems. Magistrates' courts in each state handle minor offenses without a jury. District or county courts hear criminal and some civil cases. The supreme courts of each state and of the Northern Territory deal with the most important civil and criminal offenses. They also serve as courts of appeal for cases from lower courts.

THE PEOPLE

From 1950 to 2001, Australia's population increased from about 7 million to 19.2 million. Forty percent of this growth resulted from immigration, which the government actively encouraged from 1945 to 1965. Almost one-quarter of the Australian people were born overseas.

The country's population density—six people per square mile (about ten people per sq km)—is one of the lowest in the world. Yet Australia is one of the most urbanized countries. Eighty-five percent of its people live in cities and towns. Three-quarters of the urban residents are concentrated in the eight cities that serve as state and territorial capitals. The desert and semidesert regions that make up two-thirds of Australia's land area are very sparsely populated.

Ethnic Mixture

From the early 1800s until the late 1940s, all but a small number of white Australians were of British or Irish descent. Most others could trace their ancestry to northern Europe. The Aborigines, whose population had dropped below 25,000 by 1860, lived mainly in remote areas apart from white Australians.

Faced with labor shortages after World War II, Australia attempted to increase its population by paying the travel expenses of foreigners or by giving land to people willing to settle there. Half the immigrants who took advantage of the offer were British, but newcomers also arrived from Poland, the Netherlands, Germany, Italy, Greece, and other countries. The new immigration policy brought ethnic and cultural diversity to Australia.

Until recent years, few Asians made their homes in Australia. Although the gold rush of the 1850s had attracted about forty thousand Chinese to the country, few established permanent residence. From the 1880s to the early 1970s, Australia's immigration laws discriminated against Asians who wanted to move to Australia. By 1973 these laws had been abolished.

Cambodian refugees are among Australia's Asian immigrant population.

Since then Australia has accepted many Asian refugees, including large numbers from Vietnam, Laos, and Cambodia. Other refugees have come from eastern Europe, Latin America, the Middle East, and Africa. One-fourth of the immigrants admitted during the 1980s were Asians, and people of Asian backgrounds make up about 5 percent of Australia's population.

Since 1982 job shortages and economic problems have caused the country to severely restrict immigration. The government grants visas, temporary work visas, and permanent residence to close relatives of Australian citizens, investors in new industries, and skilled or specially qualified individuals. While the refugee crisis remains a controversial political issue, the country still grants thousands of humanitarian visas each year.

Go to vgsbooks.com for information about ethnic groups in Australia, including various Aboriginal cultures. You'll also find links to websites with up-to-date population figures and other statistics.

Standard of Living

Most Australians are accustomed to a high standard of living. Most families own their homes and have at least one car. Australians travel frequently, and many who live in the outback fly their own light aircraft. Australians also enjoy an informal lifestyle, and sports and outdoor barbecues are favorite forms of recreation.

Most Australians work 38 to 40 hours per week and each year receive at least four weeks' vacation and ten paid holidays. Decades of labor shortages enabled unskilled workers to command good wages. There is a growing trend among young workers to take lower-paying service jobs, such as working in stores, restaurants, and tourist hotels. But overall, Australian incomes tend to be high.

For many years, Australian men often spent little time with their families, regarding their primary role as that of wage earner. Women worked as housewives and mothers and were underrepresented in professions and in public life. More women began to enter the workforce in the 1960s, and in 1973 a law granted federal employees the right to maternity leave. Universal maternity leave for all women workers continues to be debated. In the early twenty-first century, women made up nearly 44 percent of the workforce.

Most of Australia's Aborigines have a much lower standard of living compared to the rest of the population. About two-thirds of the country's 460,000 Aborigines live in cities. Many urban Aborigines do not have adequate housing or income, and poverty leads to major health problems. Since the early 1970s, the government has stepped up its programs to provide medical care, legal aid, and economic assistance to Aboriginal groups.

Rural Aboriginal communities are found mostly in the Northern Territory, Western Australia, and Queensland. Many rural Aborigines follow their traditional religion, laws, and social organization, providing support to community members. With government backing, Aborigines develop and operate cattle

A Lesson in Ayapathu

English is the official language of Australia. But there are about 150 Aboriginal languages still spoken by the country's indigenous people. Only about 20 of those, however, still have large communities of speakers. Most Aboriginal children learn English as a first language.

Here are some words from the Ayapathu language. Ayapathu is spoken by the native people of the Cape York Peninsula in northern Queensland.

hair	aalman
nose	kaa
toe	tha'u puku
kangaroo	kuja
dog	ku'a
mountain	yoyko
sky, cloud	yuuwa
river	wa'awa
three	ko'ele
east	kaawa
mother	paba
father	papa
little sister	wileny
cousin	moyerre
young man	uchan

stations (ranches). Other programs have helped Aborigines set up small businesses and regain some of the rural lands that once belonged to them.

In recent years, Aborigines have begun to reassert great pride in their traditional culture. Aboriginal arts thrive in written and oral literature, dance, art, music, film, poetry, and drama. Many Aborigines find an outlet for creative expression at *corroborees*, social gatherings that are highlighted by music and dancing.

Aboriginal woman

Reconciliation has become an important issue for both Aboriginal and white Australians. Many feel it is important to acknowledge past crimes against Aborigines, to heal and move forward as a united country. The acknowledgment is important to Aborigines, but what reconciliation will mean to their communities in terms of equal opportunities and better treatment remains to be seen.

Education

School attendance is compulsory for Australian children between the ages of six and fifteen (sixteen in Tasmania). Most children, however, begin attending preschool at age four or five. Two-thirds of Australia's youth attend public schools, and the remainder enroll in private schools sponsored by Protestant or Roman Catholic churches.

Secondary schooling begins after six or seven years of primary education and can continue for six more years. Many Australian students leave school after four years of secondary education. Students who complete the final two years of secondary studies can qualify for entry to one of the country's forty colleges and universities. Australia has a 99.5 percent literacy rate.

Children in the remote regions of the outback are taught through a system called Schools of the Air. Because these children would have to

Students study together in the library of a tiny Northern Territory school.

travel too far to school every day, they study at home instead. Teachers communicate directly with students over two-way radios, and lessons are given and completed through the mail, computers, television, and fax machines.

Most Aboriginal children attend the same schools as other Australian children. In recognition of the Aborigines' unique cultural heritage, however, the government offers them special preschool programs, advisers, and courses. In areas with many Aborigines, courses are taught in both an Aboriginal language and English.

Health and Social Services

Private doctors and medical personnel provide most of the health care in Australia, but Medicare, the nation's health insurance program, covers most medical expenses. The government funds Medicare through income tax collections. Treatment in public hospitals is free, and private insurance programs cover charges in private facilities. Australia's life expectancy of 80 years reflects the nation's high standard of health care.

Before the 1920s, Australians living in sparsely populated regions of the continent had no access to regular medical care. A Presbyterian minister, John Flynn, realized that pregnancies, appendicitis, even fevers, could be fatal in the outback, with no doctors or hospitals nearby. The Reverend Flynn saw two-way radios and small airplanes as the answer. In 1928 the first Flying Doctor base was established in Queensland. Later named the Royal Flying Doctor Service, it operates twelve base stations. Its staff provides medical aid, routine care, and emergency evacuations to an area about two-thirds the size of the United States.

The government funds a variety of other social services. The elderly, the long-term ill or handicapped, and single parents receive pensions (regular income payments). Financial assistance is also extended to the unemployed, the sick, those with special needs, and low-income families. All families with children receive an allowance for each child under sixteen and each full-time student under twenty-five, regardless of the family's earnings.

Traditionally, Australians have looked to their government to provide assistance in periods of economic need. At times during the 1970s and 1980s, the level of unemployment was high, and many Australians were "on the dole," as dependence on government payments is called. Continuing unemployment and inflation posed economic problems for the people of Australia into the 1990s. The Howard government has introduced economic reforms to counter those problems.

CULTURAL LIFE

Cultural life in Australia reflects the diversity of its people and the uniqueness of the land. Early immigrants brought many European cultural forms—religion, painting, traditional music, food, and wine. Australia retains its European-descended heritage and outlook. Like the indigenous Aborigines, though, the European settlers were also inspired by the continent's unique natural beauty and ruggedness. The strong oral and storytelling traditions of Ireland, England, and the Aborigines fused to create their own expression of Australian life in literature, the arts, music, and movies. Continuing immigration from other parts of the world introduces more variety to all aspects of the culture.

Nature continues to play its large part in Australian life. Relying on warm, sunny weather and a great variety of terrains, from beaches to mountains, popular culture emphasizes leisure activities and outdoor recreation. Barbecues, bush picnics, and a wide variety of sports are mainstays of life in Australia.

Religion, Holidays, and Festivals

Although many Australians do not attend church regularly, more than 70 percent are Christian. Some 22 percent belong to the Anglican Church, and 27 percent are Roman Catholic. These denominations reflect Australia's settler history, with its large percentage of English and Irish immigrants. Other Christians belong to the Uniting Church, the Baptist Church, Eastern Orthodox faiths, and the Lutheran Church. Later immigration is reflected in the largest non-Christian religions, including Islam, Judaism, and Buddhism.

Among indigenous people across Australia, there are diverse practices but very similar beliefs. Aborigines believe the world and people were created long ago in the Dreamtime. Aborigines' ancestor spirits came to earth not only as humans but as all nature—the land, plants, and animals. Ties to the ancestor spirits are maintained in totem animals and sacred places, an important reason why land rights and the environment concern Aborigines.

"At the going down of the sun / and in the evening / we will remember them." These lines were written for Anzac Day services, in memory of Australian and New Zealand soldiers killed during World War I. **Anzac Day** is a public holiday and includes parades and processions.

Anzac Day, April 25, is generally considered Australia's national holiday. It commemorates the day Australian troops landed in Gallipoli, Turkey, during World War I. Australia also observes traditional holidays such as Christmas, Easter, and New Year's Day. States celebrate their own holidays, such as Australia Day in Sydney and Proclamation Day in South Australia.

There are about 1,300 festivals and fairs a year in Australia. Residents and tourists can sample food and wine, see ethnic dancing, hear music, watch parades and fireworks, learn about history and heritage, catch comedy acts, or just enjoy spending time outdoors. Most large cities sponsor annual events. Sydney Festival, held every January, is Australia's largest annual cultural event. Melbourne has held its Moomba Waterfest since the 1950s, highlighting waterskiing, a yacht race, and extreme sports in addition to a carnival, concerts, and fireworks. The Perth International Arts Festival, also established in the 1950s, attracts 500,000 visitors a year. The Festival of Darwin celebrates the city's mix of European and Aboriginal cultures, with a parade, concerts, sports, and dances. Cities and many smaller towns also host annual festivals and movie and music gatherings with local and international performers.

Literature

Australia's natural environment and history have inspired a large literary output. Nineteenth-century authors chronicled the country's prison-colony beginnings and natural habitats. Marcus Clarke depicted prison life in his 1874 novel, *For the Term of His Natural Life.* In 1888 Thomas Alexander Browne described a gang of outlaws in *Robbery Under Arms.*

Other writers began to express their sense of nationalism, of being Australian. The farmer Joseph Furphy used the diary form to express Australian attitudes in *Such Is the Life,* published in 1903. Songs and stories about adventures in the bush and outback became popular in this period.

Some novelists wrote about what it was like not to fit into Australian society. In 1910 Ethel Richardson, writing under the name Henry Handel Richardson, published the novel *The Getting of Wisdom,* depicting life for a strong-willed Australian girl who cannot adjust to her proper young ladies' school. Richardson later gained fame abroad for her trilogy, *The Fortunes of Richard Mahoney,* about an Irish doctor who hated Australian life. Patrick White, who won the 1973 Nobel Prize for literature, wrote *Voss,* the story of a German who attempts to explore the interior of Australia.

Australia has produced many other highly regarded writers. Among the best known are Elizabeth Joley, Christina Stead, Morris West, and Michael Wilding. Colleen McCullough's novel *The Thorn Birds* introduced millions of readers throughout the world to the Australian outback. Peter Carey's *Oscar and Lucinda* won the 1988 Booker Prize and was later made into an acclaimed film. His historical novel *The True History of the Kelly Gang*

How to Speak Australian

Australians have their own way of expressing themselves, even down to the words they use. When Europeans first settled in Australia, they found a whole new world of plants, animals, and land formations they had never seen. Not knowing what to call these new things, they picked up words they heard the Aborigines using. In other cases, they used English and Irish terms in new ways. The result is a common vocabulary known as "Strine."

anklebiters: children
apples, she'll be: everything will turn out okay. "No worries. She'll be apples again."
Aussie: (pronounced "Ozzie") Australian
billabong: watering hole in the outback
bloke: an Australian male. To call someone a "good bloke" is a compliment.
dinkum, fair dinkum: true, real, genuine. "I'm a fair dinkum Aussie."
g'day: hello!
good onya: good for you, well done
mate: a friend
Never Never: the outback
no worries!: "No problem" or "Yes, I'll do it."
Oz: Australia
reckon: think
too right!: definitely!
walkabout: a walk in the outback lasting an indefinite amount of time

also won the Booker Prize in 2001. Thomas Keneally, famous as the author of *Schindler's List,* has written several historical novels set in Australia. Among nonfiction authors, Germaine Greer, born and educated in Australia, is well known for her feminist essays. Robert Hughes is best known for his history of colonial Australia, *The Fatal Shore.*

Germaine Greer

Australia has also been home to many notable poets. In the late nineteenth century, Mary Hannay Foott, Charles Harpur, and Henry Lawson reflected Australia's growing sense of itself as a nation. None were more popular, though, than the ballads of Andrew Barton "Banjo" Paterson. "Waltzing Matilda," which he wrote in 1917, is still the unofficial national song. Paterson's best-known poems and ballads are contained in *The Man from Snowy River.*

The Jindyworobaks were a group of writers in the 1930s who also tried to develop a distinctive national poetry. Although the Jindyworobaks were white Australians, they believed that Aboriginal ideas and culture had to be incorporated into any expression of Australian life. Contemporary poets have included more personal and universal themes in their works, yet their Australian identity remains important. These poets include John Trantor, Martin Johnson, Bernard Cohen, Gig Ryan, Antigone Kefala, Les Murray, and Pam Brown.

A Sunburnt Country

Dorothea Mackellar is the author of one of Australia's best-known poems, "My Country." Born into a prosperous Sydney family, Mackellar wrote "My Country" when she was nineteen. Generations of Australian schoolchildren have learned the verses, like this often-quoted one:

I love a sunburnt country,
A land of sweeping plains,
Of rugged mountain ranges,
Of droughts and flooding rains;
I love her far horizons,
I love her jewel sea,
Her beauty and her terror,
The wide brown land for me!

Art, Movies, and Music

Australian painters in the late nineteenth century were strongly influenced by French painters. Frederick McCubbin and Tom Roberts used the impressionist style to depict Australian frontier life. Artists in the twentieth century include Sir William Dobell, known for his portraits, and Sidney Nolan, who used themes from Australian folklore for his dreamlike paintings. George Russell Drysdale and Frederick

Williams painted landscapes from the outback. In recent years, John Olsen and Brett Whiteley have gained international reputations.

Aboriginal art is among the oldest in the world. Aborigines were producing bark and rock paintings and engravings twenty thousand years ago. Those techniques are still used to express cultural tradition. Other Aboriginal artists, many of whom live in cities, use a variety of techniques to reflect their life in modern Australia.

Australian filmmakers produce about twenty-five feature-length films each year, and many are distributed overseas. The country's famous movie stars include Errol Flynn, Paul Hogan, Judy Davis, and Mel Gibson. More recently, younger Australian actors have become Hollywood fixtures, among them Nicole Kidman, Cate Blanchett, Heath Ledger, Russell Crowe, and Hugh Jackman. Films from or about Australia have found an audience worldwide. Blockbusters such as *Crocodile Dundee* have made their mark, but many Australian films are smaller, independent hits such as *Strictly Ballroom, Muriel's Wedding*, and *The Dish*.

Born in New Zealand but raised in Australia, film actor Russell Crowe has won numerous awards for his dramatic performances, including an Academy Award in 2001.

Pop and dance music diva **Kylie Minogue** first burst onto the international music scene in the 1980s with her cover version of the Motown hit "The Locomotion." Although she remained popular in Australia and Britain, she didn't make an international comeback until the early twenty-first century.

Australia has also made significant contributions to the international music scene. Australian-born musicians include sopranos Nellie Melba and Joan Sutherland, guitarist John Williams, and composer Percy Grainger. In the 1980s, Australian pop and rock artists enjoyed a surge of fame. Olivia Newton-John, Rick Springfield, the Bee Gees, Air Supply, AC/DC, INXS, Midnight Oil, and Men at Work all had regular hit songs. At opposite ends of the popular music spectrum, Australians Paul Kelly and Kylie Minogue have enjoyed years of steady acclaim in their genres. Silverchair, Natalie Imbruglia, and Kasey Chambers also continue to gain popularity at home and abroad.

If you'd like to learn more about Australian culture, visit vgsbooks.com, where you'll find links to recipes, photographs, basic Aboriginal words, Aussie slang, and more.

Food

Much of the Australian diet was inherited from English and Irish settlers. Meat and potatoes were staples, and most spices beyond salt and pepper were considered exotic. "Bush tucker," as the settlers' food was called, was simple and filling but not very interesting. Fifty years of immigration, however, has changed Australian cuisine completely. Australians enjoy eating out, and diners in large cities can choose from a variety of restaurants—Chinese, Thai, Italian, Greek, French,

Japanese, and more. Many Australians also enjoy cooking themselves, especially barbecues. Cooking schools and television programs have both gained popularity.

Australia's many types of climate provide for a variety of native ingredients. Beef and mutton are still widely used, but seafood is very common, too. In the outback, exotic meats include crocodile, emu, kangaroo, and buffalo. Fresh-food markets in major cities sell strawberries, apples, citrus fruits, coffee beans, cheeses, and honeys, all grown in the country.

To go along with their diverse food, Australians enjoy many types of domestic wines. Australia's red and white table wines have consistently won awards at international shows. As with other agricultural products, wine grapes are grown in the fertile coastal areas. The major wine producing regions include the Barossa Valley, the Adelaide Hills, the Hunter Valley, and the Tamar Valley.

ANZAC BISCUITS

These mildly sweet cookies are named for the Australian and New Zealand Army Corps, which fought in World War I (1914–1918).

1 c. rolled oats
¾ c. unsweetened shredded coconut
¾ c. all-purpose flour
1 c. sugar
½ c. butter or margarine
1 tbsp. honey or light corn syrup
1½ tsp. baking soda
3 tbsp. boiling water
kitchen parchment paper or heavy-duty aluminum foil

1. Measure the oats, coconut, flour, and sugar into a medium-sized mixing bowl. Stir well.
2. In a small pan over medium-low heat, melt butter or margarine and stir in honey or corn syrup.
3. Place baking soda in a cup or small bowl. Pour boiling water over baking soda and stir to dissolve. Add to melted butter mixture.
4. Pour butter mixture over mixed dry ingredients. Mix well.
5. Preheat oven to 300°F. Cover two baking sheets with kitchen parchment paper or heavy-duty aluminum foil (dull side up).
6. Drop teaspoonfuls of dough 2 inches apart onto prepared baking sheets.
7. Bake 10 to 12 minutes, or until biscuits are crisp and golden brown.
8. Cool biscuits on a baking sheet for 1 to 2 minutes. Remove with a spatula and finish cooling on wire racks. When cool, store in an airtight container. Makes about 48 biscuits.

Sports and Recreation

Sports activities are an important part of Australian life. At least one-third of the Australian population is registered to participate in organized sports. Children take part in school sports from an early age, playing team and individual games like soccer, netball, and tennis. Many adults remain active in some sport, from amateur rugby leagues to golfing. Water sports, especially swimming and surfing, are also very popular. Eighty percent of Australians live within 30 miles of one of the country's 7,000 beaches. Many other Australians enjoy activities such as horseback riding, bushwalking (hiking), cycling, and fitness programs.

When they are not playing sports themselves, Australians enjoy watching games and matches. The most popular winter spectator sport is Australian rules football—a fast-moving game with long kicks and leaping catches. In recent years, an Australian team (the Kangaroos) has dominated international competition in rugby, a British game from which U.S. football is derived. Rugby league is the semi-professional version that provides teams for international matches. Rugby union, which draws fewer spectators, is played by amateurs.

Australian rules football, or "footy" as it is affectionately called by the natives, is closely related to the game of rugby.

Like rugby, **cricket** is a popular sport in Australia that has its roots in the British Isles.

Cricket is the country's most played and watched summer sport. Thousands of fans attend matches that bring teams from other Commonwealth nations to Australia. Horse racing is the nation's oldest sport, since an official race was run near Sydney in 1810. The Melbourne Cup race, perhaps Australia's best-known sporting event, has been held since 1861.

Yachting is also an important activity in Australia, and Australians have designed some of the world's fastest yachts. One of the greatest moments in Australia's sports history occurred in September 1983, when the *Australia II* won the America's Cup, the greatest prize in yachting. The Australian yacht was the first foreign craft in 132 years to defeat the U.S. entry.

With this emphasis on sports, it is not surprising that Australia produces many champion athletes. Evonne Goolagong Cawley and Margaret Smith Court are famous in the world of women's tennis. Adelaide native Lleyton Hewitt is top-ranked in men's tennis. Greg Norman was inducted into the World Golf Hall of Fame in 2001. Cathy Freeman and Nova Peris-Kneebone are both champion track stars and both are Aboriginal women. Ian Thorpe, a Sydney native, is known as one of the best freestyle swimmers in the world. At the 2000 Sydney Olympics, the "Thorpedo," as he is known, won three gold and two silver medals.

Australia's athletes and the country's enthusiasm for sports were center stage in September 2000, when Sydney hosted the Summer Olympics. Cathy Freeman lit the Olympic torch to open the Games, in an emotional ceremony. Freeman later won the gold medal in the 400-meter track race. After the Games, Sydneysiders, as city residents are called, were acclaimed for their hospitality. The Games were declared a great success.

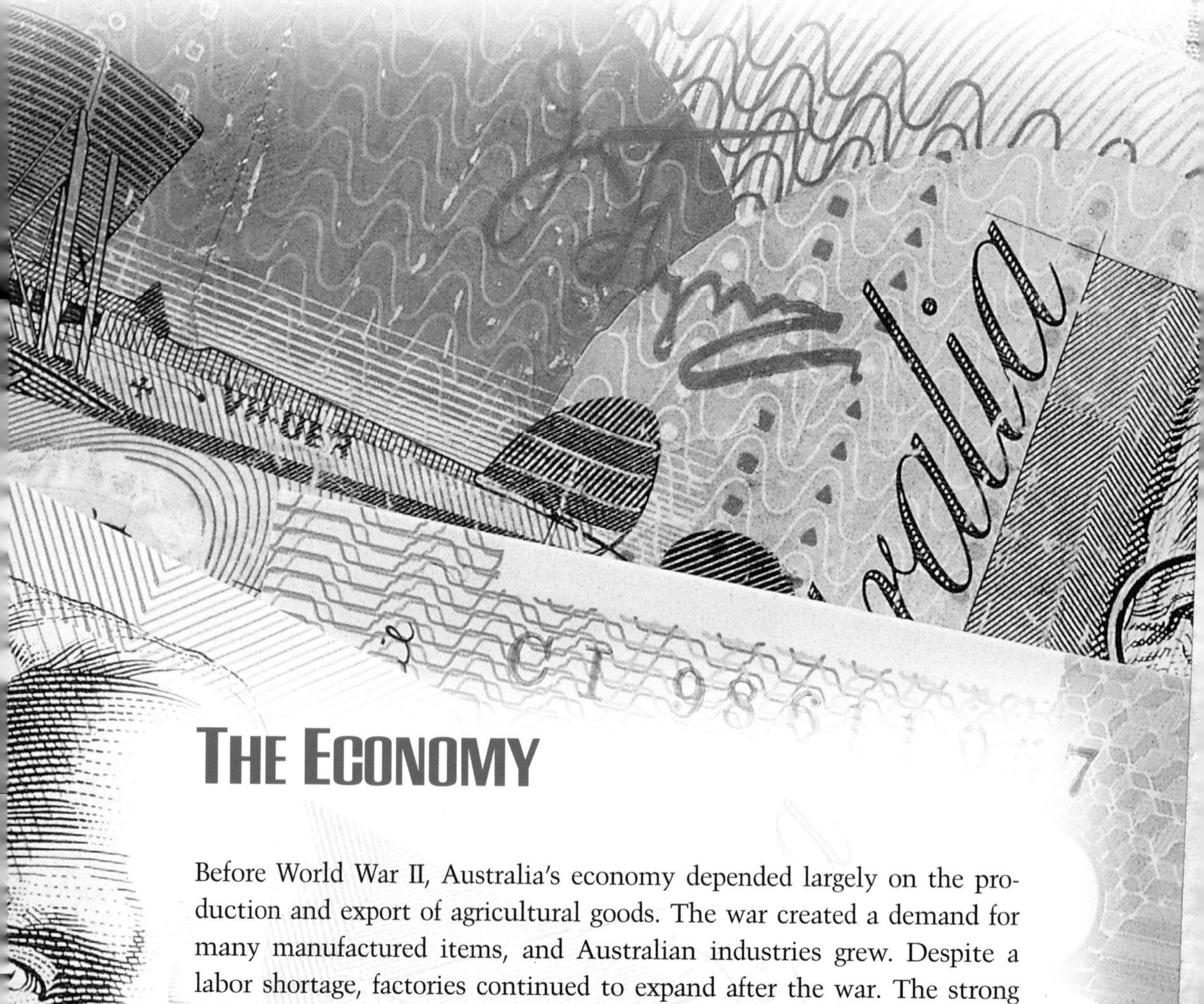

THE ECONOMY

Before World War II, Australia's economy depended largely on the production and export of agricultural goods. The war created a demand for many manufactured items, and Australian industries grew. Despite a labor shortage, factories continued to expand after the war. The strong demand for labor enabled both skilled and unskilled jobholders to earn high wages. In the 1960s, the discovery and development of mining and energy resources added to the country's prosperity. Mostly as a result of changes in the world economy, however, Australia's growth faltered from the 1970s through the mid-1990s. Foreign demand for the country's minerals and metals fell, and the market for its agricultural products shrank. Unemployment and the cost of living rose. The economy was buoyed in the late 1990s, with high growth, low inflation, and a steady stock market offsetting a fall in the Australian dollar's value.

In 2002 Australia's farmers faced the worst drought in one hundred years. Crops failed and livestock had to be sold. Bushfires resulting from the parched earth spread beyond farmlands, reaching as far as Sydney.

Experts hope that the effects of the drought will be lessened because it followed several successful farming seasons. Farmers who saved and invested during these boom times will not be wiped out by drought losses.

The crash of Asian markets, the global economic downturn after major terrorist attacks, and droughts affecting agriculture have slowed Australia's economy, but it still moves forward. In cities, in particular, business and housing investments are growing, and consumers are spending more. Inflation is expected to decline, but varying predictions about unemployment suggest that some aspects of the economy's direction remain to be seen. Economists continue to look for stability and growth in government reforms and the recovery of the global economy.

Mining and Manufacturing

Australia is rich in mineral, metal, and energy resources. The country is the world's leading exporter of coal and aluminum and a major producer of bauxite, from which aluminum is made. Australia is also

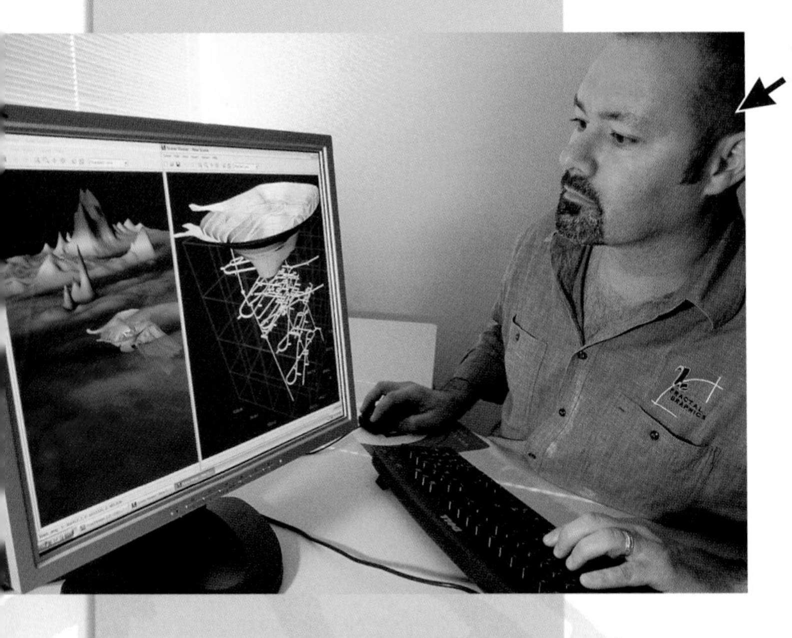

A software designer looks at a **3D computer graphic** representation of an Australian gold mine.

Mining Minds

The mining industry has produced another important commodity for Australia: mining-related inventions were the country's fifth largest mineral export in 1999. Geophysical and mining instrumentation is used to explore and develop mineral deposits. Chemistry processes help miners analyze minerals and ores as they dig. And computer software allows engineers to plan mine construction, safety needs, and production schedules on-line.

one of the top producers and exporters of iron ore, copper, nickel, gold, lead, zinc, diamonds, tin, tungsten, manganese, and zircon. Most of the world's high-quality opals come from South Australia. The country also has large reserves of uranium, used to produce nuclear energy.

Many of the country's richest mineral and ore deposits lie in remote, dry areas of the continent. Developing these resources was very expensive, since it required new roads, railways, and towns to house workers. To obtain money, mining companies turned to investors in other countries. As a result, foreigners own significant stock in Australia's mining companies. In gold mining alone, foreign control rose to 70 percent in 2002.

In 1961 geologists discovered oil 180 miles (290 km) west of Brisbane, and commercial production began in 1964. Drillers found additional deposits of oil and natural gas in the Bass Strait, the coastal waters of Western Australia, and the desert basins of both the Northern Territory and South Australia. The Bass Strait site contains immense reserves of petroleum. In 1989 Australia and Indonesia signed the Timor Gap Treaty, allowing for the cooperative exploration and development of petroleum fields under the Timor Sea. Recent political turmoil has endangered this cooperation, however.

Australia's factories make most of the consumer products the country needs. Nevertheless, the country must import most of the industrial machines and tools necessary to make these goods. Its leading manufactured products are processed foods, iron and other metals, transportation equipment, paper, chemicals, clothing, shoes, and household appliances. Sydney and Melbourne are the leading manufacturing centers.

For many years, the Australian government protected the country's manufacturers from foreign competition by enacting high tariffs. That policy changed in the mid-1980s, and most tariffs were reduced. By 2001 tariffs had dropped from 35 to 5 percent. As a result, Australian manufacturers face more competition from foreign companies in selling products to Australians. However, tariff reform allows Australia to develop its export industries, such as mining, that bring in significant income. Key markets such as Asia are also expected to grow, providing Australia with more trade opportunities.

Agriculture

Less than 5 percent of Australia's labor force are farmers, and they work on less than 60 percent of the country's land. Nevertheless, they provide almost all the food Australians need while still supplying a large export industry. Australian farms produce wool, cotton, meat, grains, sugarcane, fruit, nuts, and dairy products.

Although farms and large ranches cover 59 percent of Australia, 90 percent of the country's farmland is so dry that it can be used only as pasture. For this reason, pastoralism (livestock raising) dominates agriculture. The remaining 10 percent of the farmland is divided between crops and sown pastures. The most fertile soil lies in a 200-mile-wide (322-km) crescent-shaped band in the country's southeastern corner.

Tasmanian crop farm

Sheep and cattle dominate Australia's ranch industry. Some sheep are raised for meat, but most of their value is in wool production. Australia is the world's largest supplier of apparel wool. Seventy-five percent of all Australian sheep are merinos, a breed that

In New South Wales, a **herd of merino sheep** crowd around a water trough. Wool products are an extremely important Australian export.

produces very fine wool. Western Australia and New South Wales together raise more than half of Australia's sheep and produce half of its wool.

Cattle are raised for both meat and dairy products. Australia is also the world's largest beef exporter. In 2000 the country exported 66 percent of its beef. Half of Australia's dairy products are sent to Japan, the Philippines, Malaysia, and other countries, accounting for 16 percent of the world's dairy exports.

Wheat, the country's largest grain crop, grows in all six states. Most of the wheat is shipped abroad to China, Egypt, Russia, and Japan. Australian farmers also cultivate other grains, such as barley, oats, sorghum, rye, and millet. Sugarcane is produced in Queensland and New South Wales. Growers harvest a variety of fruits in coastal orchards and vineyards. The biggest crops are grapes, citrus fruits, apples, bananas, pears, peaches, and pineapples.

Wheat field

Australian wines sell well domestically and abroad. Exports have recently doubled. These sales, worth billions of dollars, have made Australia the world's fourth largest wine exporter. The country's chardonnays, Rieslings, and other wines are sold in Britain, the United States, New Zealand, Canada, and Germany.

Despite a gradual decline in Australia's rural population, scientific and technical

advances have pushed farm output steadily upward. In some years, however, poor weather conditions have harmed agriculture. Droughts are common, killing crops and livestock and worsening economic conditions. Fires started by lightning spread rapidly across dry scrublands, destroying whole farms in their paths. Prolonged lack of rain is not the only danger facing rural Australians. At times flash floods threaten both people and animals. The federal government plans to invest in new technology, however, to understand changes in the climate. With easy access to the Internet on home computers, farmers may be able to use this information to plan around weather patterns.

Go to vgsbooks.com for up-to-date information about Australia's economy and a site with current exchange rates where you can convert U.S. dollars to Australian dollars.

Fishing and Forestry

Many species of fish live in Australia's coastal waters, but tuna and salmon canning have the greatest commercial importance. For export Australia's small fishing industry relies primarily on shellfish, including prawns, rock lobsters, abalone, and scallops. Japan and the United States buy most of the shellfish. Smaller hauls provide whiting, snapper, barramundi, mullet, shark, and mud crabs for local buyers.

Most of Australia's forests, which cover about 21 percent of the country's land, blanket the Eastern Highlands and the moist coastal

A stack of red gum logs lies ready to be cut into planks at a Victoria lumber mill.

regions. Foresters manage the stands of trees for conservation purposes and for timber production. Most of the native trees are species of gum (eucalyptus). Some types are harvested to produce paper, furniture, and flooring, and others are valued for their gums and oils. Forestry plantations grow crops of pine—especially the Monterey pine from California—for the housing industry. Australia imports—largely from New Zealand—about one-third of the forestry products that it needs.

Transportation

Australia's National Highway system links its capital cities with more than 500,000 miles (805,000 km) of roadways. Sixteen million people drive these roads in nine million cars and motorcycles and half a million trucks. The busiest interstate connection is the Hume Highway between Sydney and Melbourne. It carries more than one thousand commercial vehicles every day, in addition to other traffic. Stuart Highway bisects the continent from Darwin in the north to Adelaide in the south. Midway along this route, near the geographic center of Australia, lies Alice Springs. From east to west, sections of the National Highway connect for more than 1,600 miles (2,575 km) to link Adelaide to Perth. Even with this extensive system, however, much of Australia's interior cannot be reached by motor vehicle. The National Highway is funded by the federal government, but states are responsible for construction and maintenance.

Australia's cities are also linked by 25,000 miles (40,234 km) of railways, and rail is one way to reach Australia's vast interior. The Great

Southern Railway runs from the country's west coast through the outback to Alice Springs. The Australian government once owned most of the country's railroads, but most are operated by private companies. Railroads also serve mining, industrial, and agricultural areas.

Qantas Airways began in the 1920s as a tiny airline serving the outback in Queensland. ("Qantas" comes from the carrier's original name—Queensland and Northern Territory Aerial Services.) This airline operates from ten Australian cities. At least twenty other international carriers also fly in and out of Australia. There are international airports in major centers such as Sydney, Brisbane, Perth, and Melbourne and connecting airports in many smaller cities. Various companies offer passenger and freight service within the country.

Freight also moves by road trains—enormous trucks with triple-size trailers. A road train can be 165 feet (50 m) long and weigh 170

In places railroads cannot reach, enormous road trains move freight across the Australian countryside.

tons (154 metric tons). Ships carry many goods in and out of Australia and move some products, such as minerals, between domestic ports.

Trade and Tourism

Agricultural and mineral products dominate Australia's export market. Minerals, fuels, and metals earn about half the country's foreign income. Coal is the most important fuel for export purposes, accounting for about 12 percent of the country's total foreign earnings. Australia sells crude oil and liquid natural gas on the world market but must import heavy crude oil to make some petroleum products. Australia sells more energy resources abroad than it imports.

Agricultural products are a major source of export earnings. As the world's leading producer of wool, Australia heavily relies on export earnings from this product. Principal crops such as wheat, sugarcane, and fruit also are important to the country's export market. Japan and the United States are Australia's leading trade partners.

Australia's major imports are vehicles, petroleum and petroleum products, office equipment, computers, industrial machinery, electrical appliances, and textiles. Japan provides most of the vehicle imports, which compete with Australia's own automobile and truck manufacturers. Saudi Arabia and Singapore furnish most of the petroleum imports. Overall, however, the United States is Australia's single largest supplier.

Through its membership in several international trade organizations, Australia seeks to promote trade by working out fair agreements on tariffs and pricing. Under the Closer Economic Relations Trade Agreement that took effect in 1983, Australia and New Zealand gradually eliminated all trade barriers. Tariffs continue to drop as Australia maintains its commitment to an open market.

Domestic and international tourism is Australia's largest industry. More than four million foreign tourists visit each year, many drawn by

A group of schoolchildren visiting Kangaroo Island takes snapshots of a koala bear. **Tourists** to Australia's many natural wonders bring in much of the country's income.

the country's great outdoors. Queensland's beaches and the Great Barrier Reef are popular with swimmers and scuba divers. Inland attractions include Uluru (Ayers Rock) and Kata Tjuta (the Olgas)—massive boulders near Alice Springs—as well as winter resorts in the Australian Alps.

Most tourists are from Japan, New Zealand, and Great Britain. Two-thirds of them arrive in Australia via Sydney. The city's harbor, opera house, and harbor bridge are among its most popular attractions. During January—when many Australians take their vacations—the city sponsors a month-long festival.

Another stopping point for vacationers is Melbourne, Australia's second largest city and Sydney's traditional rival. Though Sydney had overtaken Melbourne as the business and financial heart of the nation, Melbournians claim that their city is still the nation's social and intellectual center. Canberra and the other state capitals are also popular destinations.

The Future

In 1988, as Australia celebrated the bicentennial of British settlement, many Aborigines staged protests, saying the land originally belonged to them. Their message was heard by officials who are trying to reach compromises over land-rights issues. The future looks brighter for the Aborigines, as the government has become more sensitive to their needs.

Meanwhile, the government must continue to work for Australia's financial progress. A close alliance with Asian and Pacific nations has added to economic growth, and there seems to be little question that Australia wants open trade with other markets. A global market, with fewer government controls, also raises questions of balanced trade, environmental damage, and support for developing countries. And the effects of a rise in terrorism and political instability remain to be seen.

Amid these struggles, the Australian people are examining what it means to be Australian. Many citizens favor cutting the country's traditional ties to the British monarchy. Much of the Australian culture has roots in Britain, yet one-quarter of Australia's population was born abroad. In the years ahead, the members of this multicultural society must seek ways of successfully working and living together, while playing an increasingly important role in world politics.

C. 12,000 B.C. Tasmania separates geologically from the Australian mainland.

C. 10,000 B.C. The first humans settle in Australia.

1606 Dutch explorer Willem Jansz reaches northeastern Australia.

1642 Dutch explorer Abel Tasman sails around the continent and discovers the island of Van Dieman's Land (Tasmania).

1688 British explorer William Dampier arrives in Australia. He returns in 1699.

1770 Captain James Cook of the British Royal Navy reaches the southeastern corner of the continent.

1788 The first fleet of convict ships arrives in Botany Bay on Australia's southeast coast.

1813 Lachlan Macquarie, governor of New South Wales, encourages the first crossing of the Blue Mountains.

1829 Britain claims the western half of the continent. Captain Charles Sturt sets out to explore the country's interior.

1835 The city of Melbourne is founded.

1851 Gold is discovered at Bathurst in New South Wales. Even larger deposits are discovered in Victoria.

1897 Delegates to a national convention begin to draw up a federal constitution.

1901 Australia's six British colonies adopt a federal constitution and become one nation. Commonwealth adopts immigration laws known as the White Australia Policy.

1913 Construction begins on the planned capital city of Canberra.

1914 Australia enters World War I after Britain declares war against Germany.

1915 On April 25, Anzac forces land in Gallipoli, Turkey.

1917 Andrew "Banjo" Paterson writes "Waltzing Matilda" while visiting the outback.

1928 The first Flying Doctor base is established in Queensland.

1929 The U.S. stock market crashes, and the Great Depression begins across the world.

1932 Sydney Harbour Bridge opens.

1942 Japanese planes and submarines attack Australia, which joins forces with the United States against the threat of invasion.

1945 World War II ends.

1947 Australian government opens immigration to non-British Europeans.

1949 Australia becomes a founding member of the United Nations. Robert Menzies is elected prime minister.

1956 Melbourne hosts the Olympic Games.

1961 Geologists discover oil west of Brisbane. Commercial production on the fields begins in 1964.

1966 Australia adopts the decimal system for its currency, changing from its old system of pounds and pence.

1967 Aborigines are granted citizenship and can vote for the first time.

1972 Gough Whitlam is elected prime minister, the first Labor politician to hold the office in more than twenty-five years.

1973 Government ends White Australia policy, allowing immigration from Asia.

1974 The National Highway System begins paving existing roads and building new ones.

1988 Australia celebrates its 200th birthday with huge parties throughout the country.

1992 The Australian High Court legally recognizes that Aborigines have rights of ownership (native title) to traditional lands.

1993 The Native Title Act passes, returning some lands to Aboriginal groups and compensating others.

1999 Australians vote against a referendum making the country a republic.

2000 More than two hundred refugees drown when three boats sink off the coast of Australia. Sydney hosts the Olympic Games in September.

2001 Australia celebrates the Centenary of Federation.

2002 In January inmates at Woomera jail stage a hunger strike in protest of living conditions. On October 12, suicide bombers kill eighty-six Australians and other tourists in a resort area on the island of Bali. Australian farmers face the worst drought in a century.

COUNTRY NAME Commonwealth of Australia

AREA 2,969,909 square miles (7,692,064 sq. km)

MAIN LANDFORMS Central Lowlands, Eastern Highlands (Great Dividing Range), Great Western Plateau, the outback

HIGHEST POINT Mount Kosciusko, 7,310 feet (2,228 m) above sea level

LOWEST POINT Lake Eyre, 49 feet (15 m) above sea level

MAJOR RIVERS Darling, Goulburn, Lachlan, Murray, Murrumbidgee

ANIMALS Dingoes, emus, kangaroos, koala bears, parrots, platypus, spiny anteaters, Tasmanian devils, wombats

CAPITAL CITY Canberra

OTHER MAJOR CITIES Sydney, Melbourne, Brisbane, Perth, Adelaide, Hobart

OFFICIAL LANGUAGE English

MONETARY UNIT Australian Dollar. 100 cents = 1 dollar.

Fast Facts

Currency

AUSTRALIA'S CURRENCY

Australia's monetary unit is the Australian dollar. As part of the British Commonwealth, Australia once used currency similar to Britain's, with pounds, shillings, and pence. In 1966 the country changed to its current decimal system.

Notes, or paper money, feature portraits of famous Australians, such as social reformer Catherine Helen Spence, journalist Banjo Paterson, and Aboriginal writer David Unaipon. Notes are printed in denominations of $5, $10, $20, $50, and $100.

The obverse (or front) of all Australian coins features Queen Elizabeth II, as the reigning monarch. Reverse sides depict the country's birds and animals or its coat of arms. Coins are minted in denominations of 5, 10, 20, and 50 cents, and $1 and $2.

Australia's flag reflects both its history and its uniqueness. Called the Commonwealth Blue Ensign, the flag features a Union Jack (Britain's flag), one seven-pointed star, and an arrangement of five other stars.

The Union Jack represents Australia's history as a British colony. The seven-pointed star is called the Star of Federation or the Commonwealth Star. Its points symbolize Australia's six states and combined territories. The five stars represent the Southern Cross constellation. The constellation has always guided sailors through the Southern Hemisphere and has become emblematic of Australia's geographic position.

The Commonwealth Blue Ensign was selected after a 1901 public competition of more than thirty thousand designs. Although the winning design was announced in 1903, and Australians began flying the flag, it was another half-century before it was officially adopted, with the Flags Act of 1953.

Flag

National Anthem

Until 1974 the Australian national anthem was Great Britain's "God Save the Queen." After a government opinion poll, "Advance Australia Fair" was selected as the new anthem. The first verse is printed below.

Australians all let us rejoice,
For we are young and free,
We've golden soil and wealth for toil,
Our home is girt by sea;
Our land abounds in nature's gifts,
Of beauty rich and rare,
In history's page, let every stage,
Advance Australia Fair.
In joyful strains then let us sing,
Advance Australia Fair.

For a link to a site where you can listen to Australia's national anthem, "Advance Australia Fair," go to vgsbooks.com.

EVONNE GOOLAGONG CAWLEY (b. 1951) This Australian tennis champion was one of the most popular players of the 1970s. A singles player, Cawley won seven Grand Slam tournaments, Wimbledon twice, the Australian Open five times, and the French Open once. Born in Griffith in New South Wales, she is part Aboriginal in ancestry. She retired from tennis in 1983.

VICTOR CHANG (1936–1991) Born in Shanghai to Australian-born Chinese parents, Victor Chang became a cardiac surgeon after graduating from Sydney University. A pioneer of the modern era of heart transplantation, he lobbied for funding for the research and development of an artificial heart valve and, later, an artificial heart. In 1986 Chang was awarded a Companion of the Order of Australia.

CAROLINE CHISHOLM (1808–1877) The daughter of an English farmer, Caroline Chisholm first arrived in Australia as the wife of a British officer. Through her work on behalf of women and children, she became known as "the emigrant's friend." She helped not only those women coming into Australia but also those left behind in England. She worked for the protection of single women during their immigrant voyages and then provided housing for them once they arrived. She also arranged family reunions for men working in Australia who could not afford to send for their wives and children.

CATHY FREEMAN (b. 1973) This popular Australian athlete was born in Queensland. She is the first Aborigine to win an Olympic gold medal, for the 400-meter race in Sydney in 2000. She was chosen to light the Olympic flame at the opening ceremonies of the Sydney Olympics. She was also named Young Australian of the Year in 1990 and Australian of the Year in 1998, the only person to win both titles.

PEARL GIBBS (1901–1983) An Aborgine born in Sydney, Gibbs began working at age 16 as a domestic servant in wealthy homes. Later she became active in the Aboriginal rights movement, one of the few women to do so at the time. She worked the rest of her life for political equality and the repeal of discriminatory laws. In 2001 an Australian newspaper writing about the hundredth anniversary of her birth called Gibbs "the mother of reconciliation."

STEVE IRWIN (b. 1962) The Crocodile Hunter has become familiar to thousands of television viewers. The son of game park keepers, Irwin grew up feeding and caring for wild animals. As the director of the Australia Zoo in Queensland, Irwin does not so much hunt crocs as rescue, relocate, and care for them. He has also introduced his TV audiences to many native Australian reptiles and other fauna. Irwin was born in the city of Victoria and grew up in Queensland.

NED KELLY (1854–1880) The country's most famous bushranger, or bandit, was born near Melbourne to Irish immigrant parents. Facing poverty, Kelly began stealing from the rich at an early age. He graduated from horse thievery to bank robbery and was hanged at age twenty-five. Although some consider him a violent criminal, Kelly became a folk hero to others, a symbol of rebellion against injustice.

EDDIE MABO (1936–1992) Eddie Mabo was a Torres Strait Islander, one of a group of Aborigines whose land was considered part of Queensland. While working as a gardener at James Cook University, Mabo sat in on classes and read college library books. He was shocked to learn that early British settlers to Australia had declared the entire country *terra nullius,* or "no one's land." If it were no one's land, the British were free to take it. But Mabo knew Aborigines had lived in Australia for thousands of years. He organized and led a battle to change the law of *terra nullius.* He died just months before the High Court of Australia overturned *terra nullius,* but the historic Mabo decision was named after him.

SIR ROBERT GORDON MENZIES (1894–1978) Menzies served as prime minister longer than any other person in Australian history. He helped found the Liberal Party and dominated Australian political life during the 1950s and 1960s. He was born in a small town in Victoria, where his father owned a grocery store. Always a good student, Menzies graduated from law school with top honors. After practicing law for several years, he was elected to Victoria's state parliament, then went on to federal politics.

DAME JOAN SUTHERLAND (b. 1926) Born in Sydney, Australia's opera diva remembers first singing with her mother as a child. After debuting in England in the 1950s, she went on to become, in many people's opinion, the best soprano of the twentieth century. She was made a Commander of the Order of Australia in 1975, in recognition of her contribution to opera.

IAN THORPE (b. 1982) Known as "the Thorpedo," Thorpe is well on his way to becoming one of Australia's greatest swimmers. He has already broken over twenty world records. At his first Olympic Games, in his hometown of Sydney in 2000, Thorpe won three gold and two silver medals. He was allergic to chlorine as a child but took up swimming anyway after watching his older sister compete. He won his first medal at the age of eight.

PETER WEIR (b. 1944) Born and raised in Sydney, Weir studied art and law before becoming an acclaimed film director. His first film to gain attention outside his native country was *Picnic at Hanging Rock*, in 1975. He then directed Mel Gibson in *Gallipoli* and *The Year of Living Dangerously*. Many of the films that followed were commercial and critical successes: *Witness*, *The Mosquito Coast*, *The Dead Poet's Society*, and *The Truman Show*.

Bungle Bungles In Purnululu National Park in Western Australia, the sandstone rock formations of the Bungle Bungles Range rise up like enormous beehives. Silica and algae color the mounds with orange and black stripes. The range is an estimated 350 million years old. Aborigines have used the range for thousands of years, but few white people even knew about it until twenty years ago. Bungle Bungles can only be reached by four-wheel drive vehicles. Scenic flights in helicopters or small airplanes also give a bird's-eye view of the park.

Great Barrier Reef The largest coral deposit in the world sits off the northeastern coast of Queensland, in the Coral Sea. The reef extends 1,250 miles (2,012 km) from north to south and is composed of more than one thousand islands. Entire ecosystems form around the reefs making them excellent spots for scuba diving and snorkeling. The Great Barrier is particularly popular with divers and swimmers, as the seawaters are warm and clear year-round.

Kakadu National Park This natural reserve in the Northern Territory has been continuously inhabited for forty thousand years. Its wide range of ecosystems include forests, plateaus, floodplains, and waterways, each supporting a variety of plants and animals. Flowers, mammals, birds, butterflies, reptiles, and fish all flourish. Many of these species are found only in Kakadu. Thousands of years of Aboriginal art and traditional sites also enrich the park.

Sydney Harbour Two of Australia's most famous urban symbols can be seen in Sydney Harbour: the opera house and the bridge. The Sydney Opera House, which took almost fifteen years to build, looks from a distance like the white sails of a group of boats. It is a complex of almost one thousand rooms, where symphony, ballet, dance, theater, and opera performances are staged. Sydney Harbour Bridge, known locally as "the Coat Hanger," is one of the longest single-span bridges in the world. It connects the central business district and North Sydney. The waterfront surrounding Sydney Harbour is also famous for its parks, historical sites, shops, and restaurants.

Uluru The world's largest rock sits in the center of Australia, in Kata Tjuta National Park. Uluru, also known as Ayers Rock, is a monolith, or a single block of solid stone. It rises 1,043 feet (318 m) above the desert floor and is almost 5 miles (8 km) around the base. Some geologists believe it is the top of a mountain that extends 3 miles (4.8 km) beneath the surface. The rock is made of sandstone and feldspar, which give it its red color. That color changes as the daylight and atmospheric conditions shift. It is especially stunning at sunrise and sunset. Uluru is sacred to the Anangu Aborigines, who believe their ancestral spirits still occupy the surrounding area. Several caves in Uluru contain ancient Aboriginal rock paintings.

Aborigines: the first human inhabitants of Australia. Their descendants retain many unique traditions and beliefs.

Anzac: the Australian and New Zealand Army Corps. A volunteer military force during World War I, the Anzac troops are best known for their bravery against terrible odds in the battle for Gallipoli, Turkey. Anzac Day, in commemoration of the battle, is celebrated throughout Australia on April 25.

artesian: a type of well that draws water from underground reserves. Those reserves are under great pressure, so the water comes to the surface naturally. Much of Australia's Central Lowlands lie above an artesian basin, from which livestock are watered.

bush: a large unpopulated area, usually scrubland. In Australia the term also generally means any wilderness. The bush figures large in Australian folklore and has given rise to many other terms, such as bushwalking, bushranger, and bush tucker.

didgeridoos: an Aboriginal musical instrument made from naturally hollowed tree limbs, usually eucalyptus. Didgeridoos are played at rituals and celebrations.

marsupial: an animal that develops its offspring in a pouch. More than half of Australia's 230 native animal species are marsupials, including kangaroos.

monotreme: a mammal that lays eggs but nurses its young

outback: a vast, sparsely populated area in the interior of Australia. The continent is one of the world's flattest and driest landmasses, so much of its interior consists of plains and deserts.

parliament: a lawmaking assembly. In Australia the Federal Parliament is continuously in operation. Members are returned (elected) from their voting districts, and the leader of the majority party becomes prime minister, the head of parliament.

playa: the floor of a desert lake

privatization: the practice of changing a business or industry from public to private ownership

scrubland: an area covered with low trees and small plants and shrubs

territory: a subdivision of a country, organized under its own legislature but still dependent on the federal government. Two of Australia's eight political subdivisions are territories.

transportation: the practice of sending criminals from the British Isles to Britain's colonies to work as slave labor. England shipped 150,000 convicts to Australia before the practice of transportation ended in the mid-1800s.

***Australian Bureau of Statistics 2002.* (March 6, 2003)**
Website: <http://www.abs.gov.au> **(January 14, 2003)**
ABS is Australia's official statistical organization. Its website provides census statistics and special articles. In addition to general population tables, studies are grouped under categories such as health care, immigration, mining, and tourism. More in-depth analyses are available for a fee.

***Australian Tourist Commission.* N.d.**
Website: <http://www.australia.com> **(March 6, 2003)**
Facts, features, destinations, festivals, and other travel details are listed by state or city. An interactive section, photo gallery, and downloadable fact sheets provide more in-depth information about Australia.

Darian-Smith, Kate. *Australia and Oceania.* Austin, TX: Steck-Vaughn Co., 1997.
This title begins with a geological and geographical study of the Australian continent and its surroundings. It also includes information on the climate, history, government, and people.

***Economist.com.* (March 6, 2003)**
Website: <http:www.economist.com/> **(January 7, 2003)**
This on-line version of the 150-year-old journal provides analysis of world business and current affairs, including Australia, as well as background information and country surveys. There is also a searchable archive of all the *Economist's* articles back to June 1997.

***The Europa World Year Book 2002.* London: Europa Publications Limited, 2002.**
This annual publication covers Australia's recent history, economy, and government. It provides a wealth of statistics on population, employment, trade, and more.

Kaplan, Gisela. *Women in Society: Australia.* New York: Marshall Cavendish Corporation, 1993.
This title is a study of Australian women in general, their history, and their changing role in a diverse culture. Interspersed are short biographies of early reformers, athletes, artists, and politicians.

Macintyre, Stuart. *A Concise History of Australia.* Cambridge, England: Cambridge University Press, 1999.
Australia's history is examined with a particular focus on politics and economics.

Nile, Richard, and Christian Clerk. *A Cultural Atlas of Australia, New Zealand, and the South Pacific.* Abingdon, Oxfordshire, England: Andromeda Oxford Limited, 1996.
The atlas covers general geological and zoological information as well as Australia's cultural history.

***Smh.com.* (March 6, 2003)**
Website: <http//www.smh.com/> **(March 5, 2003)**
The *Sydney Morning Herald* on-line features articles about Australian politics, general news stories, finance, sports, and entertainment.

Smith, Rolf M. *National Geographic Traveler: Australia.* Washington, D.C.: National Geographic Society, 1999.
This edition is a full-color study of Australia with many interesting details.

Stein, R. Conrad. *Cities of the World: Sydney.* New York: Children's Press, 1998.
This study of Australia's largest and busiest city also features some general information about the country.

***Theage.com.* (March 6, 2003)**
Website: <http://www.theage.com/> **(January 6, 2003)**
Melbourne's daily newspaper, the *Age*, offers news, politics, and sports.

Turner, Barry, ed. *The Statesman's Yearbook: The Politics, Cultures, and Economics of the World 2003.* New York: Palgrave Macmillan Ltd., 2002.
A general survey of Australian history and government is provided, with tables from the Australian Bureau of Statistics.

***Washingtonpost.com.* (March 6, 2003)**
Website: <http:/www.washingtonpost.com/> **(January 6, 2003)**
The *Post*'s International section provides daily news from around the world, including Australia. It also features on-line archives dating back to 1977, from which articles can be downloaded for a small fee.

The Australian Consulate General New York
Website: <http://australianyc.org>
Learn about Australian business, culture, government, and tourism from this overseas post of the Department of Foreign Affairs and Trade.

Bartlett, Anne. *The Aboriginal Peoples of Australia.* Minneapolis: Lerner Publications Company, 2002.
Take an in-depth look at the history, cultural traditions, and modern life of Australia's first people.

Bryson, Bill. *In a Sunburned Country.* New York: Broadway Books, 2000.
American travel essayist Bill Bryson writes with affection and humor about a country "mostly empty and a long way away."

Conway, Jill Ker. *The Road from Coorain.* New York: Vintage Books, 1990.
Conway's autobiography begins with her childhood in the Australian outback, recounting the details of her family life and their relationship to the rugged land. It ends when Conway leaves the country as a young woman bound for Harvard University in the United States.

Finley, Carol. *Aboriginal Art of Australia.* Minneapolis: Lerner Publications Company, 1999.
Learn about Australia's Aborigines from their art—traditions of painting and engraving dating back thousands of years.

Franklin, Miles. *My Brilliant Career.* 1901. Reprint, Pymble, Australia: HarperCollins Publishers, 2002.
Franklin published *My Brilliant Career*, her first novel, in 1901, when she was just 22 years old. Its heroine Sybylla is a funny, intelligent, rebellious young woman who longs for something beyond her life in rural Australia.

Germaine, Elizabeth, and Ann L. Burckhardt. *Cooking the Australian Way.* Minneapolis: Lerner Publications Company, 2004.
From lamingtons to sunshine salad to Pavlova, learn to make the best-loved foods of Australia.

Habegger, Larry, ed. *Australia: True Stories of Life Down Under.* San Francisco: Traveler's Tales, Inc., 2000.
A host of famous travel writers, including Paul Theroux and Jan Morris, contribute to this explanation of life down under.

Hughes, Robert. *The Fatal Shore: The Epic of Australia's Founding.* New York: Vintage Books, 1998.
Critic and author Robert Hughes, an Australian expatriate himself, tackles the long and extraordinary history of colonial Australia.

Parliament of Australia
Website: <http://www.aph.gov.au/> **(February 17, 2003)**
Australia's federal legislative body offers information on who's who, how Parliament works, what bills are being discussed, and other governmental information.

Richardson, Henry Handel. *The Getting of Wisdom.* McLean, Virginia: Indypublish.com, 2002.
Richardson's 1910 novel recounts the story of Laura, a young girl sent from the freedom of the Australian countryside to a proper young ladies school.

Shute, Nevil. *A Town Like Alice.* New York: Ballantine Books, 1991.
Shute's World War II novel tells the story of Jean, a woman captured by the Japanese army in Malaysia. In the prison camp, she befriends Joe, an Australian prisoner-of-war.

Sutton, Peter, ed. *Dreamings: The Art of the Aboriginal Australians.* New York: George Braziller, a division of W.W. Norton & Sons, 1997.
Australian art critic Robert Hughes called this volume "the best short introduction to the Aboriginal world view now in print."

vgsbooks.com
Website: <http://www.vgsbooks.com>
Visit vgsbooks.com, the homepage of the Visual Geography Series®. You can get linked to all sorts of useful on-line information, including geographical, historical, demographic, cultural, and economic websites. The vgsbooks.com site is a great resource for late-breaking news and statistics.

Index

Aborgines, 4, 5, 7, 16, 24, 26, 30, 31, 33, 34, 35, 41, 44, 45, 46, 50, 65; culture of, 20–22; standard of living of, 43
acacia. *See* wattle
Adelaide, 18, 62
agriculture, 10, 18–19, 23, 24, 56, 57, 59–61, 63; cattle industry, 19, 59
Alice Springs, 19, 62, 63, 65
Anzac (Australians and New Zealand Army Corps), 31, 48
ANZUS (Australia, New Zealand, United States), 32
art, 50–51; Aboriginal, 51
Asia, 4, 7, 8, 16, 36, 37, 59
Australia: boundaries, location, and size, 8; climate, 14, 53, 61; coast, 10–12; currency, 68; flag, 23, 69; flora and fauna, 15–16; geographic regions, 9; independence, 5, 29, 30; maps, 6, 11; national anthem, 69; states, 8, 38; territories, 8, 38; topography, 9–12
Australian Alps, 13, 65
Australian Capital Territory (ACT), 19, 38
Ayers Rock. *See* Uluru

Bass Strait, 9, 58
Botany Bay, 22
Brisbane, 18, 27, 63
British monarchy, 34, 37, 65

Canberra, 19, 29, 34, 65
Cape York Peninsula, 12
cities, 17–19, 40
climate, 53, 61
Commonwealth Coat of Arms, 39
constitution, 29, 37
convict colony, 5, 17, 18, 19, 22, 24, 27, 29, 49
Cook, Captain James, 22
Coral Sea, Battle of the, 32
corroborees, 44
cultural life, 46–56; Aboriginal, 44

Darwin, 19, 32, 62
deserts, 10, 15, 40
didgeridoo, 21
diversity, 7, 33, 36, 40–41, 46, 47, 53, 65

Eastern Highlands, 9, 12, 61
economy, 5, 7, 27, 34, 35, 56–65; problems with, 7, 29, 31, 33, 34, 42, 56; reform of, 35, 45
education, 44–45
emu, 16, 39
energy, 5, 56
eucalyptus. *See* gum tree
Eyre, Lake, 10, 13

film, 51
flag, 23, 69
flora and fauna, 15–16
food, 52–53
forestry, 61–62

Gallipoli, 31, 48
government, 5, 7, 19, 26, 29, 30, 33, 34, 37–39, 45, 62, 65
Great Artesian Basin, 9, 10, 13
Great Australian Bight, 9, 10
Great Barrier Reef, 12, 65
Great Britain, 4, 5, 18, 22, 23, 25, 27, 28, 30, 31, 32, 64
Great Dividing Range, 12, 25
Great Southern Railway, 62–63
Great Western Plateau, 9, 10
gum tree, 14, 62

Hargreaves, Edward, 27
health care, 45
history: 20–34; British settlement, 22–26; Dutch explorers, 22; exploration and expansion, 26–27; first Australians, 20–22; gold rush, 27–29, 30, 40; self-government, 28; unification and independence, 29, 30; World War I, 31; World War II, 32
holidays, 48
Howard, John, 35, 36, 37

immigration, 5, 7, 28, 29, 30, 31, 32, 33, 36, 41, 42, 46, 47, 52. *See also* refugees
Indonesia, 36, 59
industry, 5 18, 19, 35, 56, 63

Jansz, Willem, 22
Japan, 32, 36, 61
Jindyworobaks, 50

kangaroo, 15, 26, 39
Kata Tjuta (the Olgas), 65
Keating, Paul, 35, 36
Kerr, John, 34
koala, 15, 64–65
kookaburra, 16

labor laws, 29
Labor Party, 32, 33, 35
labor strikes, 33, 34
language, 43, 45, 49
Liberal-National coalition, 35
Liberal Party, 32, 33, 34
literature, 48–50

Macquarie, Lachlan, 25
manufacturing, 18, 19, 31, 33, 59
marsupials, 15
Melbourne, 18, 27, 29, 59, 62, 63, 65
Menzies, Robert, 32
military, 19. *See also* Anzac (Australians and New Zealand Army Corps)
mining, 5,17, 19, 28, 33, 34, 56, 57–59, 63
music, 52

National Party, 33
natural resources, 17, 18, 57–59, 64
New South Wales, 8, 10, 12, 17, 19, 23, 27, 28, 38, 60
New South Wales Corps, 24, 25
New Zealand, 31, 32, 62
Norfolk Island, 24
Northern Territory, 8, 10, 19, 30, 38, 39, 43, 58

oil, 58–59
Olympic Games, 55
outback, 9, 13, 18, 19, 27, 42, 44, 45, 49, 53, 63

Parliament, 19, 30, 33, 34, 39
Perth, 17, 18, 27, 62, 63
political parties, 29, 33, 34, 35
population, 17, 29, 31, 40
Port Adelaide, 19

Queensland, 8, 10, 12, 17, 18, 26, 28, 30, 38, 43, 45, 60, 63, 65

racial prejudice, 29, 30, 41. *See also* White Australia policy
rain forest, 15
reconciliation, 35, 44
recreation, 42, 46, 65
refugees, 33, 36, 42
religion, 4, 27, 47
rivers, 10, 13, 14, 27
Royal Flying Doctor Service, 45
Rum Corps, the, 25

Schools of the Air, 44–45
seas and oceans, 4, 8–9, 13, 18, 19, 59
sheep industry, 10, 59
Snowy Mountains, 13
social services, 35, 45
South Australia, 8, 10, 13, 27, 28, 38, 48, 58
Southeast Asia, 20, 32, 33, 36
sports, 54–55
standard of living, 5, 33, 34, 42–44
Sydney, 17, 18, 23, 24, 25, 26, 48, 55, 59, 62, 63, 65

Tasman, Abel, 22
Tasmania, 8, 9, 12, 14, 19, 21, 22, 27, 28, 38, 44
Tasmanian devil, 15
terrorism, 37, 57, 65
tourism, 18, 19, 35, 64–65
trade, 36, 56, 59, 64, 65
transportation, 62–64

Uluru (Ayers Rock), 19, 34, 65
United Nations, 32

Victoria, 8, 10, 12, 18, 27, 28, 38
Vietnam War, 33

water resources, 13–14, 27
wattle, 15, 39
Western Australia, 8,10, 17, 27, 28, 33, 38, 43, 58
White Australia policy, 30, 33
Whitlam, Gough, 33
wine, 53, 60
women, 33, 43
World War I, 48
World War II, 41, 56

Captions for photos appearing on cover and chapter openers:

Cover: The Devil's Marbles are one of many unusual rock formations in Australia's outback.

pp. 4–5 A view across Sydney Harbour gives a breathtaking panorama of the world-famous Sydney Opera House and the Harbour Bridge.

pp. 8–9 Uluru, or Ayers Rock, is the largest solid rock on earth. Thousands of tourists flock to this natural wonder each year.

pp. 20–21 This rock painting was done on a cave wall in Northern Australia by ancient Aborigines.

pp. 40–41 A vineyard owner proudly displays his crop. Australian wines have become popular with afficionados worldwide.

pp. 46–47 A hiker peers into the canyon at the southern end of Major Mitchell Plateau in Victoria. Outdoor activities such as mountain biking, rock climbing, camping, and hiking are very popular with Australians.

pp. 56–57 A collection of Australian dollar bills

Photo Acknowledgments
© Bill Bachman, pp. 4–5, 12 (top), 13, 14, 15 (top), 16 (both), 17, 19, 20–21, 40–41, 44 (both), 46–47, 54–55, 54 (bottom), 56–57, 58, 59, 60–61, 60 (bottom), 62, 63; Digital Cartographics, pp. 6, 11; © Betty Crowell, pp. 8–9, 18, 39; © Rich Kirchner, pp. 10, 12 (bottom), 15 (bottom), 64–65; National Library of Australia, pp. 22, 24 (both), 25, 28, 30, 31, 34; © Hulton | Archive by Getty Images, p. 23; © CORBIS, p. 26; © Reuters NewMedia Inc./CORBIS, pp. 37, 52; © Pacific Pictures/John Penisten, p. 38; © Howard Davies/CORBIS, p. 42; © AFP/ CORBIS, pp. 48, 51; © UPPA/Zuma Press, p. 50; © Michael McGraw, p. 68.

Front cover: © Bill Bachman. Back cover: NASA